Advance Praise
Aquaponic Gardening

Aquaponic Gardening is an excellent primer for anyone considering home-scale aquaculture. Whatever your location or methods, the information should prove invaluable. Fish are within reach! — Peter Bane, publisher, *Permaculture Activist* and author, *The Permaculture Handbook*

This is a comprehensive handbook on how to grow real food, so meticulously documented, that failure is not an option. — Jeff Edwards, president, Progressive Gardening Trade Association

I have always wanted to figure out how to do sustainable aquaculture in the context of my home garden. Finally I've got the book to help me do it. — Paul Greenberg, author, *Four Fish: The Future of the Last Wild Food*

This is a delightful book to read! I've been involved with hydroponics and aquaculture for 30 years and still learned from reading this very thorough how-to book.
—Henry A. Robitaille, PhD, former general manager, The Land Exhibit, Epcot Center

Learning how to garden through the creation of a completely balanced ecosystem is now clearly understandable, even to inexperienced gardeners.
—Michael C. Metallo, President and CEO, National Gardening Association

Sylvia Bernstein has provided the "aquapons of the world" with a clear, impassioned, and elegant "Bible" to spread the good news about aquaponics.
— James J. Godsil, cofounder, Sweet Water Organics, Sweet Water Foundation

Now the thousands of people who are discovering aquaponics every day have a resource for moving from the dream to the step-by-step reality of raising fish and food in their homes, yards, and even businesses. — John Thompson, AeroGrow International, Inc.

This book is a vital resource for urban homesteaders.
— Sundari Kraft, author, *The Complete Idiot's Guide to Urban Homesteading*

The science is so well explained, it is easily understood. I am ready to start. I love this book!
— Jeff Lowenfels, author, *Teaming With Microbes*

I believe that home-scale aquaponics will become as common as the backyard chicken coop as we move toward a regenerative future that has made food security a priority. — Marco Chung-Shu Lam, Permaculture teacher, Environmental Studies Adjunct Faculty, Naropa University

The book we've all been waiting for.... a truly comprehensive guide to all things aquaponic.
— Charlie Price, founder, Aquaponics UK

Sylvia masterfully lays out the art of giving balance to an ecosystem of flora and fauna.
— Britta Riley, founder, Windowfarms.com

For those of you who want to grow fish I definitely recommend this book as a simplified method of constructing and operating an aquaponic garden.
— Dr. Howard M. Resh, author, *Hydroponic Food Production*

If you want to garden aquaponically, this is the one source that will guide you from start to finish while also taking you on a wonderful trip through Sylvia's own personal aquaponic journey. — Gina Cavaliero , managing director, Green Acre Organics / Aquaponics Enterprises, Inc.

Every time I enter Sylvia's aquaponic greenhouse, a powerful sense of inspired well-being envelops me almost instantly, and after reading *Aquaponic Gardening*, I understand why.
— Dr. Virginia F. Gurley MD, MPH, founder, Auraviva

This book is easy to read and is packed with information that will be very useful to the beginner and advanced aquaponics practitioner alike.
— Murray Hallam, founder, Practical Aquaponics

When it comes to the emerging field of aquaponics, Sylvia Bernstein is one of those inspired innovators you need to pay close attention to. — Thomas Frey, DaVinci Institute

Sylvia Bernstein has taken an immensely detailed, complex, and sometimes contradictory body of knowledge, broken it down into easily understood bites of information, and infused it with her deep passion for this emerging field. — Susanne Friend, owner, Friendly Aquaponics

A practical, easy-to-follow guide that takes the mystery out of aquaponics. Now everyone can grow their own food even if they do not have a green thumb.
— Ann Forsthoefel, former executive director, Portland Farmers Market

It might take a little bit of time for the general public to catch up with us and other "early adopters," but when they do (and they will) this book is going to be the top book recommended to them by all who really know what they're talking about.
— Jesse Hull and Molly Stanek, Imagine Aquaponics

My wish is that Sylvia's revolutionary "how to" aquaponic wisdom becomes an adopted approach to food cultivation. — Matt McMullen, director, Facilities Management and Sustainability, University Corporation for Atmospheric Research

Sylvia Bernstein's passion for aquaponics, and personal stake in the subject make this book an essential read for anyone interested in the concept of sustainably produced food.
— Marijke Peters, producer, Earth Beat, Radio Netherlands Worldwide

SYLVIA BERNSTEIN

AQUAPONIC GARDENING

A STEP-BY-STEP GUIDE to RAISING VEGETABLES and FISH TOGETHER

NEW SOCIETY PUBLISHERS

Cover design by Diane McIntosh. Cover Images: Water splash, © iStock (Okea);
Arugula, © iStock (elzeva); Tilapia, © iStock (Daniel Loiselle); Insets - Peppers, © iStock
(David Gomez); Zucchini, © iStock (Denis Pogostin); Bok Choy, © iStock (MentalArt);
Tomatoes, © iStock (Dan Driedger)

Printed in Canada. Second printing November 2011.

Paperback ISBN: 978-0-86571-701-5
eISBN: 978-1-55092-489-3

Inquiries regarding requests to reprint all or part of *Aquaponic Gardening* should
be addressed to New Society Publishers at the address below.

To order directly from the publishers, please call toll-free (North America)
1-800-567-6772, or order online at www.newsociety.com

Any other inquiries can be directed by mail to:

New Society Publishers
P.O. Box 189, Gabriola Island, BC V0R 1X0, Canada
(250) 247-9737

Library and Archives Canada Cataloguing in Publication

Bernstein, Sylvia
    Aquaponic gardening : a step-by-step guide to raising vegetables and fish
together / Sylvia Bernstein.

Includes index.
ISBN 978-0-86571-701-5

    1. Aquaculture. 2. Hydroponics. I. Title.

SB126.5.B47 2011   635'.048   C2011-904699-7

New Society Publishers' mission is to publish books that contribute in fundamental ways to
building an ecologically sustainable and just society, and to do so with the least possible impact
on the environment, in a manner that models this vision. We are committed to doing this not
just through education, but through action. Our printed, bound books are printed on Forest
Stewardship Council-certified acid-free paper that is **100% post-consumer recycled** (100%
old growth forest-free), processed chlorine free, and printed with vegetable-based, low-VOC
inks, with covers produced using FSC-certified stock. New Society also works to reduce its
carbon footprint, and purchases carbon offsets based on an annual audit to ensure a carbon
neutral footprint. For further information, or to browse our full list of books and purchase
securely, visit our website at: www.newsociety.com

NEW SOCIETY PUBLISHERS

For Alan
My husband, parenting partner, business partner,
editor and best friend

and

For the Aquaponic Gardening Community,
my inspiration

Books for Wiser Living
recommended by *Mother Earth News*

Today, more than ever before, our society is seeking ways to live more conscientiously. To help bring you the very best inspiration and information about greener, more sustainable lifestyles, *Mother Earth News* is recommending select New Society Publishers' books to its readers. For more than 30 years, *Mother Earth* has been North America's "Original Guide to Living Wisely," creating books and magazines for people with a passion for self-reliance and a desire to live in harmony with nature. Across the countryside and in our cities, New Society Publishers and *Mother Earth* are leading the way to a wiser, more sustainable world.

Contents

Acknowledgments

The very first person I need to thank is my incredible husband and business partner, Alan. When I signed up to write this book for New Society in early November, it was with the understanding that the book was due in early March. Four months to write a 200-page book is a very short amount of time and we knew it was going to be a very intense time for us. When I took on this project he took on much of my load with the business and our home without complaint. On top of that he has been my editor and coach throughout the process. He has read and meticulously edited every chapter, sometimes twice. He has done whatever has been necessary to clear the way for me to meet target dates, from screening my calls to picking up Chinese food for dinner for the third time in a week. When we married twenty years ago, I knew he would be a great husband and father but I also got the best business partner and editor I can imagine. Some women just get lucky.

The next person I want to thank is Dr. Wilson Lennard, who runs Aquaponic Solutions in Australia. Dr. Lennard views himself and his PhD in aquaponics as a resource to the aquaponics community and has been nothing but generous in sharing his knowledge and time with me on this project. In many ways, the genesis for this book was a document that he and I created together for the Aquaponic Gardening Community called "The Aquaponic Gardening Rules of Thumb." When he heard I was writing this book, he immediately offered to help in any way I wished. He has reviewed, and sometimes re-reviewed, all of the more scientific chapters of this book and contributed enormously to each of them. All he asked for in return is an acknowledgement. Clearly that is the least I can do.

I'd also like to give credit to my other Australian aquaponics friend and mentor, Murray Hallam of Practical Aquaponics. I consider Murray to be the top expert in the world in media-based aquaponic gardening systems for the home. His excellent videos *Aquaponics Made Easy* and *Aquaponics Secrets* were really the first attempt to take the chatter of the forums and create an understandable education program for the home aquaponics gardener. In a newly emerging field like aquaponics, it can be challenging to separate the good information from the bad. I always know I can turn to Murray for good, time-tested, practical guidance I can trust.

In writing this book I have also relied on the contributions of and conversations with others who have played important roles in developing the new world of home aquaponics: Travis Hughey, the author of the Barrel-ponics® manual; Rebecca Nelson and John Pade, who taught the first aquaponics workshop I ever attended, wrote the first book on aquaponics, and publish the *Aquaponics Journal*; Joel Malcolm, who publishes *Backyard Aquaponics* magazine and runs the Backyard Aquaponics forum; and Paula Speraneo of S&S AquaFarms, who runs the aquaponics email list.

Then there is the Aquaponic Gardening Community, from whom I have learned so much. Media-based home aquaponics has been developed not by corporations or universities, but by individuals around the world tinkering and experimenting, and then reporting the results online. The Aquaponic Gardening Community is a central worldwide hub for the free exchange of information and experiences about aquaponics. Every day, people are in there helping each other out, posting photos and videos, and slowly but surely advancing the shared knowledge base of aquaponics.

Within this community there are some members to whom I owe a particular debt of gratitude. First, those who shared their personal aquaponic stories: Amy Crawford, Tawnya Sawyer, Raychel Watkins, Andrea Keene and Bill Hahn. Next, the members who lent their personal expertise in a subject matter article: Nate Storey, Kellen Weissenbach, Affnan, Kobus Jooste and Rob Torcellini. Then there is everyone who offered their answer to the question "What does aquaponics mean to you?" that populated the quotes at the top of each chapter: Sahib Punjabi, Rick Op, Daniel E Murphy, Ted J. Hill, Molly Stanek, Michelle Silva, Darryl Hinson, Paul Letby, Dan Brown,

Teddy Malen, Jeffrey Mays, Jim Knott, Richard Wyman, Gina Cavaliero and Tonya Penick. Finally, to everyone who has ever contributed to a discussion in the forum — thank you.

I also owe a debt of gratitude to Tom Alexander for the beautiful foreword he wrote to this book. As a long-time admirer of both the magazine he published (*The Growing Edge*) and his personal writings, I was thrilled when Tom agreed to tackle the foreword. That he did so with such enthusiasm and depth was a rewarding bonus.

And my gratitude to Kim Leszczynski, my long time friend and graphic designer, who cares deeply about quality and doesn't stop working on an project just because the budget has run out.

Finally, I'd like to thank my publisher, New Society. If you hadn't believed in the power of aquaponics and been convinced that this book needed to be published it might have never been written. You have given respect and beauty to a subject that I hold dear, and I thank you for that.

Foreword

By Tom Alexander

The United States is blessed with an abundance of fertile soil in most states that support traditional soil-based agriculture, producing harvests of all types of crops, both for consumption within the USA and for export. In my thirty years of reporting and publishing articles on agriculture around the world, I saw firsthand that other places are not so fortunate. Countries like Australia, New Zealand, Israel and Holland rely on their not-so-fertile soil to act like a foundation base for acreages of hydroponic greenhouses and aquaponic systems to produce enough vegetables and fish to feed their people.

Now, with the pressure to produce more food to feed an ever-increasing world population, even countries with abundant areas of fertile soil are looking at both hydroponics and aquaponics to produce fish or food crops both in a faster growth cycle and in more volume in a given space. With the correct inputs, hydroponics and aquaponics systems both fit those demands.

Health-conscious consumers also want an increasing quality of food. "Locavores" and "foodies" are terms that didn't exist ten years ago. But now, all areas of the developed world have large locavore foodie populations along with a growing Slow Food movement that demands locally grown, fresh produce in the meals they eat, both at home or in restaurants. It matters not whether those tasty food items are grown on a local farm or in a home's basement or backyard; aquaponics fills the bill for locavore foodies' demands for freshly harvested, locally grown food.

Aquaponics can be used to raise fish and fresh produce at any scale, from very large commercial systems to very small personal setups and everything in between. Whatever the size, all aquaponics systems use the same concepts and technology. The common limitations for personal use are space, know-how and motivation. Anyone thinking about throwing the fish in the tank and planting a few seedlings in a hydroponic system while waiting for a successful harvest in a few weeks is in for a rude awakening of both crop failure and system failure. Both the fish and the plants growing in their respective systems need regular visual and technical monitoring. If adjustments need to be made, they need to be made immediately. Soil acts as a buffer to plants when deficiencies occur. In aquaponics, both the plants and their roots are in direct contact with the water solution and react fast in a negative manner to any deficiencies or imbalances. This is where the book you are holding in your hands comes into the equation. After reading it, you will have all the information you need to master the technology and become successful in aquaponics.

I first met Sylvia Bernstein when she was working as the Vice President of Marketing and Product Development for AeroGrow International promoting their flagship product, the AeroGarden. It was the first truly "plug-and-play," attractive, tabletop hydroponic unit for the kitchen that would grow fresh herbs and greens for "foodies" to use in their culinary creations. It reminded me of an inkjet printer and was as simple to use as one. Sylvia would give presentations on the unit and its aeroponic technology at progressive gardening conferences and trade shows I attended. I was impressed with her knowledge of the technology, her enthusiasm in explaining it to her audience and the quality of the information she shared. Sylvia was so convinced with the success of the hydroponic technology and believed in it so much that she broke out on her own, starting an internet site that has become one of the top sites to go to for everything aquaponics-related. Both the beginner and advanced aquaponic grower can and will learn something from her site.

After doing much research, collecting and publishing a lot of information on aquaponics, Sylvia and her husband designed and manufactured a backyard aquaponics system that is simple yet effective in producing great quantities of both fish and food for the home grower.

Hydroponics is efficient in its use of water (by recirculating/recycling it within the closed-loop system) and in the time it takes to grow finished produce.

Lettuce for example only needs 26 to 30 days to mature, compared to the 45 to 48 days it takes in soil-based systems. Aquaponics not only has those benefits but also brings fish into the equation. The fish obviously offer a new harvest of a different crop and also provide organic food source for the hydroponic crops. By recirculating the fish-waste water to the food crops, the fish waste is used up by the plants as a nutrient. This in effect "polishes" the water clean of the fish waste and it is then recirculated back to the fish tank. Most aquaponic growers use fish that will be a food source, such as tilapia, but some are also raising species such as koi and goldfish, which are used in ornamental landscape ponds. It all depends on the type of fish the local market demands. For backyard production it depends on the type of fish desired on the plate in the dining room. Tilapia and barramundi are two common species raised in aquaponics but I have also seen systems raising trout, bass in Australia and even freshwater shrimp in New Zealand.

Sylvia first investigated using aquaponics to raise fish and produce on a commercial scale for consumers in the Denver, Colorado metro area. Investigated is the key word here. She realized, after thoroughly researching commercial aquaponics, that it was not the right fit for her. However, she acquired a huge quantity of information, links and personal contacts in her research. This lead to her starting a business model around home aquaponics so that she could share what she learned with others. This book is a key part of her business model and I predict it will become one of the "bibles" of aquaponics.

Aquaponic Gardening: A Step-by-Step Guide to Raising Vegetables and Fish Together is a book targeted at the personal, backyard or basement grower. However, within its covers is valuable information that both the small personal-use grower and the commercial grower can use for a successful harvest of both fish and vegetables.

It is written with Sylvia's personal accounts of her trial and error in using aquaponics at her own house. Trial and error is a substantial part of the human-interest angle of this book. Sylvia shares what has worked for her and what has not. After you consume your first fish and vegetables harvested from your system, I hope you will thank Sylvia, at least in your thoughts, for what she has done to help you in your new aquaponic adventure. This book is the first building block to your success.

After publishing *The Growing Edge* magazine for more than twenty years and hearing feedback from experienced growers who had learned something from an article they had read, I believe people experienced in aquaponics can also learn something from a book like this one. Even if they only learn one thing, that one thing could save them hundreds, even thousands of dollars by making their operation more efficient and ultimately more successful.

In the hydroponic and aquaponic industries, I have found that some people are very secretive about their techniques. However, the vast majority of people in these industries are very open to sharing what they have learned, while taking pride in being of service to newbies and helping a fairly new agricultural industry succeed in the years to come. Sylvia is one of the latter types of people.

Reading about aquaponics can get a person excited about the potential of using the technology to raise fish and grow vegetables. Seeing a working aquaponics operation firsthand will motivate a person beyond excitement to try it himself or herself. Having a consultant to coach someone who is new to aquaponics is a luxury that most people don't have. Sylvia's book can be your on-call 24–7 aquaponics consultant, as close as your bookshelf! I hope you use it frequently.

TOM ALEXANDER was publisher of the print magazine *The Growing Edge* from 1989 until 2009. *The Growing Edge* continues to report on all aspects of progressive gardening and agriculture, including greenhouses, hydroponics and aquaponics, on their free, web-only site, www.growingedge.com.

Preface

"Nature has all the answers. What was your question?"

— Howard T. Odum, noted ecologist

The aquaponics epiphany

The rain was a gift. I had set aside that entire Saturday in early April to do yard work, but instead was searching for something to do inside. As it happened, my then 14-year-old son also had no plans, and my husband and daughter were out of town. Hmmm. What to do? Clearly something together would be best, but where was the overlap in our current interests? Then it hit me. I remembered my longtime buddy at AeroGrow, John, had been for trying for months to get me to come over to see his basement aquaponic system. Fish and crawdads growing plants in a basement might be interesting. The added bonus of seeing their new baby chicks sealed the deal. We got in the car and drove off without realizing that our lives were about to be changed forever.

The AeroGarden.

AeroGrow International, Inc.

I admit I was skeptical. John and I were both part of the original founding team at AeroGrow, the makers of the AeroGarden. The AeroGarden is a small, countertop-sized hydroponic garden about the size of a toaster oven. It grew plants year-round, indoors, without dirt or weeds. It was the first product that really took hydroponics out from its hiding place in closets and basements and brought it to the mass markets and the Average Joes. John and I were the only gardeners of the five original founders, and later on the executive team. We often felt that we had an unspoken, but profound, responsibility to the gardening world. Why? We wanted to not only make sure that this very special product got to market, but that it made it in a way that got gardeners excited. "They are ruining our system!" we often secretly complained, behind a closed door in one or the other of our offices, or on a walk if it got really bad. But by working together as a united, "green" front, we generally prevailed and managed to launch a product of which we are both extremely proud.

We were born within hours of being one year apart in age and were often teased for being more like sister and brother than co-workers. I love him like family, but like any siblings we have marked differences in our personalities that sometimes caused misunderstandings and battles. John is a dreamer, an inventor, a "ready, fire, aim" kind of guy. I am more studied, measured and skeptical. I need proof. John had been trying to convince me for months that he really was growing plants with just the water from fish, but I figured that this was just another one of his wild dreams.

So, with this as background, you can see why I was dubious when I approached his home that rainy Saturday. I had occasionally heard of aquaponics over the years through the hydroponic trade magazines. But I had always dismissed it as more of a desire by the environmental fringe to change the fundamentals of hydroponic growing than a viable reality. While I am not a scientist by education or title, I know a lot about growing plants. I am a longtime traditional dirt gardener with experience spanning four yards over four states. I joined AeroGrow in 2003 and soon set up and managed the Grow Lab and Plant Nursery. We developed the hydroponic plant nutrients, a pH-buffering system and other seed-kit technologies that are the basis for several of the patents that list me among the inventors. I then became the Director of Plant Products and assumed the responsibilities for the rest of the seed kit product line. By the time I left AeroGrow in October of 2009 I was the VP of Marketing and Product Development. Why did I leave such an interesting job? In part, it was time to move on. AeroGrow had become a very different place than the company I had joined so many years ago. The main reason, however, was to pursue what had become a true passion — aquaponics.

When my son and I walked into the basement of John's rural ranch house, we were greeted with the sounds of baby chicks scratching on their newspaper-lined flooring and water flowing among the grow beds. The room was well lit and warm from the plant growing lights. The air smelled moist and fresh. The plants I saw were healthy and huge and the fish were active and obviously hungry as John tossed in a handful of food. He was excited to show us his setup and to debunk my skepticism. He was right. Aquaponics works!

When I saw that basement setup, I was immediately convinced that aquaponics was going to become a very important growing technology. I concluded that it solves the problem of expensive, and often unsafe, chemical fertilizers in hydroponics. It solves the problem of waste removal in aquaculture. It solves the problem of excess water use in traditional agriculture. And for the backyard gardener, it solves the problems of weeds, under- and over-watering, fertilizing and back strain.

Since that rainy April day, I've dedicated my life to learning all I can about aquaponics and spreading the word about this amazing way to grow plants. In December 2009, I started the Aquaponic Gardening Blog (aquaponicgardeningblog.com) to write about my personal journey through aquaponics. Topics have ranged from practical advice on seed starting and grow bed depth to musings about organic certification and visits with fellow aquaponic addicts. In January 2010, I started the Aquaponic Gardening Community site (aquaponicscommunity.com). It has become a thriving meeting place for worldwide, round-the-clock conversations about aquaponics. This community, and other aquaponics communities and forum sites around the world, are an incredible source of shared learning and support for this burgeoning new growing technique and industry. Without online communities I wouldn't be writing this book.

Because it was unrealistic to think that I could live off of blogging and running a niche community, that winter my husband and I also started a company called The Aquaponic Source (theaquaponicsource.com). It brings aquaponics education, community and products together under one roof. We have designed an aquaponics system called AquaBundance and have many ideas for future products. I teach, I speak and I have produced a video called *Aquaponics Explained*. My goal is to spread the word about aquaponics to any willing audience and to take it from an obscure gardening technique with just a few converts into a worldwide movement.

About this book

Home-scale aquaponic gardeners have evolved from the early tinkerers setting up systems in their backyards and basements. They learned from the academic work that was going on in North Carolina and the Virgin Islands, and then focused on making systems that were simpler and cheaper to build and operate. They wanted systems that could grow a wide variety of crops, not just salad greens and tilapia. They wanted to use recycled materials and off-the-shelf parts. They wanted the least amount of monitoring and fuss possible.

They found each other, and started talking online.

Probably one of the earliest examples was in West Plains, Missouri, in the early 1990s. Tom and Paula Speraneo created a successful media-based aquaponics farm called S&S AquaFarm and subsequently wrote a guide for others to follow what they had learned. More importantly, they also started an email list to start a worldwide conversation about aquaponic gardening.

In the early 2000's Joel Malcom, an engineer from Perth, Australia started looking into aquaponics and found very little published information, but he did find the Speraneo's list-serv. He started experimenting with his own backyard system, wrote a book about his experience called *Backyard Aquaponics*, and founded a company by the same name. Now Joel also runs the world's largest aquaponics forum site and is the publisher of *Backyard Aquaponics* Magazine, in addition to his very successful aquaponics systems and supplies business.

At the same time that Joel started his aquaponics adventure, fellow Australian Murray Hallam heard about aquaponics, and Joel, and struck up a relationship with him. Although Murray also had an aquaponic system business on the other side of the country, the two men collaborated and learned from each other. Murray's company, Practical Aquaponics, also sells aquaponic systems and supplies and he also runs a large forum site. Murray

Ecofilms Australia

Murray Hallam, President
of Practical Aquaponics.

is now probably best known worldwide for his entertaining educational Aquaponic video series.

In large part due to the efforts of these two men, home-based or backyard aquaponics has quickly become an accepted part of the Australian gardening scene. But these efforts have certainly been fostered by the match between aquaponics and some particulars of the Australian environment. Most of Australia enjoys year-round growing conditions. This enables aquaponic systems to be set up unprotected, without fear of winter freezing and bacteria die-off. Australia has also been experiencing one of their worst droughts in recorded history so the water-conserving benefit of aquaponics is especially appealing there. Finally, in the recent devastating floods in Queensland, aquaponics again proved to be uniquely adapted to Australia. While grocery store produce aisles were picked over, aquaponic gardeners were picking fresh veggies from their raised grow beds.

Back in North America, aquaponics took a different, two-pronged path. First, university efforts in the Virgin Islands, and to a lesser extent in a few other places, were targeting commercial applications that weren't appropriate for the backyard gardener. The other efforts were largely led by folks like Travis Hughey (Barrel-ponics®) and Will Allen (Growing Power), whose aquaponic designs were created with an eye toward solving urban or third-world food problems rather than an optimal growing experience for a North American backyard gardener.

In October 2010, after six months of learning about aquaponics and all its benefits, and spending many, many hours on the Backyard Aquaponics forum, I quit my job at AeroGrow to focus on building an industry around backyard aquaponics, American style. Since that time, I have started a blog, a company and a community site, all focused on what I refer to as "aquaponic gardening" — media-based aquaponics for growing vegetables and fish at home in a variety of climates.

One of the things I've learned since starting these endeavors is that while aquaponics excites more and more people every day, those people are not finding the reliable information they need to get started and grow successfully. Forums and community sites are tremendous, critical resources but you need a lot of time and patience to wade through the thousands of accumulated posts and often chatty or acrimonious exchanges. You also need to be able to separate the wheat from the chaff, the good information from the bad. This is not always easy in a new technology where the information providers are usually everyday folks growing a huge variety of fish and plants under a huge variety of conditions. Where do the universal truths lie?

One day I got this message in my Aquaponic Gardening Community inbox:

> "We need something that people who are starting up could hang onto. I realize that there is no set way but I think what we need to do is tell the new people what we do and does it work. I consider myself educated but I am not an engineer or a lot of other things and I need advice. I don't need arguments over which advice is correct. As I get older I find I need less complicated explanations."

I decided it was time to write this book.

What this book is, and isn't, about

My aim for this book is to provide a comprehensive guide for successful home aquaponic gardening. With it, you now have all the information you need in order to grow using aquaponic techniques. You have guidelines on how you can create your own system, or how to shop intelligently for a system kit. You will know how to start your system, when to add fish, how many to add, and how to take care of them. You will know the same for the plants. You will have a precise set of guidelines for monitoring and maintaining your system.

I started the process of writing this book in 2009 when I began to write the Aquaponic Gardening Blog, and wrote a series of articles for *Backyard Aquaponics, Growing Edge* and *Urban Garden* magazines. I wrote the real backbone, however, in November 2010, when I created a set of Aquaponic Gardening Rules of Thumb in collaboration with Australian Dr. Wilson Lennard.

Dr. Lennard earned one of the few PhDs in aquaponics in the world in 2006. After that he designed, constructed and managed Minnamurra Aquaponics, Australia's first truly commercial-scale aquaponic system. Dr. Lennard writes extensively on aquaponics for both scientific and trade journals, and currently consults worldwide through his company, Aquaponic Solutions.

The guidelines Dr. Lennard and I developed have also been reviewed and endorsed by Murray Hallam of Practical Aquaponics and vetted by the Aquaponic Gardening Community and my blog community. While there are exceptions to almost every Rule, I can guarantee you that if you follow them as they are written you will have a successful aquaponic gardening experience.

These Rules are now available through my website, community site and blog, and are reprinted in the back of this book. You will find that at the end of each chapter, I have also listed the subset of the Rules that were detailed in that particular chapter.

Media-based aquaponics

This book is entirely focused on media-based systems, with a brief discussion of vertical and hybrid systems. Why media-based and not raft or nutrient film technique (NFT)? For two reasons: solids filtration and planting flexibility.

In a raft system (also known as deep-water culture or DWC), the plants are started in a media cube, then that cube

Green Acre Organics farm, Brooksville, Florida, owned and operated by Tonya Penick and Gina Cavaliero.

is anchored into a hole in a floating board (typically Styrofoam) or "raft". This raft is then placed in a channel of oxygenated fish-waste water and the plant roots grow directly into that water. This system works well, until the solid fish waste starts accumulating on the plant roots and starves them of oxygen. To get around this problem, raft-based systems typically include some or all of the following additional filtering components: a mineralization tank, a degassing tank and a clarifying tank. While the addition of these three components enables you to stock your aquaponics system with more fish, the extra cost and complication of adding these to your aquaponics setup just does not make sense for a backyard gardener. Plus, a filtration system means that you need to clean out the filtrate on a daily basis and dispose of it outside the system. I'm not crazy about this for two reasons. First, who needs one more thing to clean in their lives? Second, why remove valuable sources of plant nutrient from the system if it just isn't necessary?

In a media-based system the grow bed becomes the filtration system for all the waste products. If a media system is constructed, stocked and operated as instructed in this book, the only solids removal that will need to be done will be a monthly shot of high-pressure water through your pipes and pump to knock lose any solid waste buildup inside. Otherwise there is no regular cleanout of the grow beds or fish tanks. Ever.

The second reason why media systems are more appropriate for home gardening is that there is almost no limit to the types of plants you can grow

Frog and zucchini.

Ecofilms Australia

in these systems. Raft and NFT systems have lower levels of nutrients because of the solids removal and they pose logistical constraints around a set of holes arranged on a fixed grid on a floating raft. These both conspire to limit the types of plants that grow best to smaller, nitrogen-loving plants like salad greens and herbs.

In contrast, a media-based system can grow absolutely anything that doesn't require an acidic soil (blueberries, for example, don't do well in any aquaponic system). I know a woman in Florida who has grown a banana tree. Murray Hallam has grown papaya trees.

I have grown 25-foot tomato plants, ground cherries, peppers, cucumbers, strawberries, celery, broccoli … as well as salad greens and herbs. Backyard gardeners do not want to be limited in what they can grow.

How this book is structured

I want you to think of aquaponic gardening as a technology platform, like your personal computer. We will be talking about a variety of hardware options, such as grow beds, fish tanks and plumbing components. We will also be talking about the possible software, or living, options you have for your system, i.e., the fish and the plants. By the end of this book you will be able to configure your own, customized aquaponic system based on the combination of options that make the most sense for your gardening and food growing goals. Best of all, I assure you that even though you will be creating your own customized system, it will work, and work well, if you follow the Rules of Thumbs that are woven throughout the book. No matter how you configure your system, the underlying Rules remain the same.

The chapters in this book are arranged in logical, chronological order for building a system. I recommend reading them in the order they appear in, and reading them all. Because aquaponics is an integrated ecosystem, every component needs to be present and in balance with all the other components. Unless you teach yourself how they all work together, you risk doing something that adversely affects that balance.

The members of the Aquaponic Gardening Community are present throughout this book, either explicitly or implicitly. Each chapter after the introduction is started by an experienced, or sometimes not so experienced, aquaponic gardener from the site, answering the question "What does aquaponics mean to you?" I hope you find their insights and passions as inspiring as I did.

Section 1

An introduction to aquaponics

"Here's to the crazy ones. The misfits. The rebels. The troublemakers. The round pegs in the square holes. The ones who see things differently. They're not fond of rules. And they have no respect for the status quo. You can quote them, disagree with them, glorify or vilify them. About the only thing you can't do is ignore them. Because they change things. They push the human race forward. And while some may see them as the crazy ones, we see genius. Because the people who are crazy enough to think they can change the world, are the ones who do."

— Apple Inc.

The author and her dog.

Benjamin Rasmussen

After breakfast Luna, my nine-year-old Tibetan terrier, runs to the door that leads to the back deck whenever I approach it. She knows that soon we need to go feed the fish. After several false alarms, the moment finally arrives when I open the door. She races down the stairs, banks around the corner and skids to a stop in front of the greenhouse. When I finally join her, I open the door and we are both momentarily overwhelmed by the sights, smells and sounds that greet us every morning.

The winter aquaponic garden in my greenhouse is alive in a way that the cold, still outside garden can't possibly aspire to. The warm, moist air smells slightly like freshly turned earth after a spring rain. The sound of flowing water tells of life and energy. The vibrant green plants in various stages and sizes are bursting with promise and productivity. A ladybug flies by. But the best part is the fish. I glance down at Luna, who has pulled herself up on the rim of my 300-gallon stock tank with her front paws to peer at the community within. She never tires of watching them, forever hopeful that someday, if they get just close enough, and she is just fast enough, just maybe…

Welcome to aquaponic gardening. With this book I hope to take you on a journey through an entirely different way of gardening. You will learn how to grow plants in rocks using only fish waste as the fertilizer source and bacteria and worms as the bridge between barren toxicity and harmonious fertility. It sounds simple, and in many ways it is, but it can also have a profound effect on your ability to feed yourself and those around you.

With this technique you will learn how to grow edible fish to supplement your family's diet with safe protein you raised yourself. You will learn to grow fruits, vegetables and greens using less than a tenth of the water and without the weeds of a traditional soil garden. And you will be able to grow food anywhere, without the restrictions of soil and sunlight.

Aquaponic gardening is a fascinating and enjoyable hobby, but fair warning — it can be very addictive. Yes, it is a healthy addiction, like yoga or salads, but an addiction nonetheless. For some, this means expanding from time to time to keep "the itch scratched." I've seen systems start with a 30-gallon aquarium and one small bed, then become 300 gallons and four beds. Pretty soon the addicts are raising bass and trout in a newly converted backyard pool.

My personal experience has been a tale of expansion as well. I started with a 70-gallon pond liner from Home Depot. When that sprung a leak (I didn't puncture it, I swear), I replaced it with a 120-gallon version — the fish were getting bigger and needed more room, right? Now I'm up to 120 tilapia and assorted goldfish in five tanks — four 60-gallon and one 300-gallon. Sad but true. Save yourself while you still can.

1

What is aquaponics?

So what is this crazy, addictive gardening technique? Here is one attempt at a definition:

> Aquaponics is the cultivation of fish and plants together in a constructed, recirculating ecosystem utilizing natural bacterial cycles to convert fish waste to plant nutrients. This is an environmentally friendly, natural food-growing method that harnesses the best attributes of aquaculture and hydroponics without the need to discard any water or filtrate or add chemical fertilizers.
>
> — Aquaponic Gardening Community, November 2010

The above was the result of a month-long online effort to define this thing called aquaponics. It is an excellent starting point for describing what it is that separates aquaponics from any other growing system available today. Let's look under the hood at the individual components of this definition:

1. "cultivation" — This is a system of agriculture for growing the plants and fish we want to consume, rather than a description of a wild, uncultivated environment.
2. "fish and plants together" — These four words describe the heart of aquaponics. Without fish and plants being grown together, you don't have aquaponics.

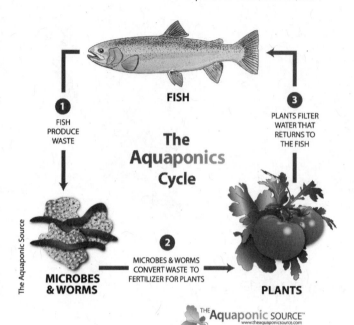

The Aquaponic Source

FISH

1 FISH PRODUCE WASTE

The Aquaponics Cycle

3 PLANTS FILTER WATER THAT RETURNS TO THE FISH

2 MICROBES & WORMS CONVERT WASTE TO FERTILIZER FOR PLANTS

MICROBES & WORMS

PLANTS

THE Aquaponic SOURCE™
www.theaquaponicsource.com

3. "ecosystem" — The dictionary defines an ecosystem as "a system formed by the interaction of a community of organisms with their environment." Aquaponics is an ecosystem of plants, fish, bacteria and worms.

4. "constructed ecosystem" — This eliminates plants being grown on the shores of a lake or pond from the definition of aquaponics. While we are centering on a notion of an ecosystem, it must be an ecosystem that is constructed for the purpose of growing fish and plants together.

5. "recirculating ecosystem" — This constructed ecosystem must also retain its water by recirculating it rather than allowing it to drain off into the water table. This is why aquaponics uses so little water compared to the systems that spawned it.

6. "utilizing natural bacterial cycles to convert fish wastes to plant nutrients" — This speaks to the key mechanism that enables aquaponics to work. Without the nitrifying bacteria that convert the fish waste into plant food, the fish would soon die in their own waste, and the plants would starve for lack of nutrition.

In other words, aquaponics is a system where plants and fish are grown together symbiotically. The waste product from the fish provides the food for the plants, and the plants in turn filter the water that goes back to the fish.

This is an environmentally friendly, natural food-growing method that harnesses the best attributes of aquaculture and hydroponics without the need to discard any water or filtrate or add chemical fertilizer.

The second part of the definition focuses on the key benefits of aquaponics and introduces the notion that it is really the combination of two other sophisticated cultivation techniques: hydroponics and aquaculture. Both of these techniques require more intervention than an aquaponics system. Aquaculture has to ensure that the waste from the fish is removed before it builds to toxic levels, or the fish will die. Hydroponics requires a constant replenishment and manual balancing of the chemical nutrients, or the plants die. By combining the two systems, aquaponics transfers much of the responsibility for reaching equilibrium between the filtration of the fish waste and the nutrient needs of the plants to Mother Nature.

The second part of the definition also asserts that in combining these two techniques, the major problems of each are solved while the major benefits are retained. That is an incredible assertion. Before we go there and decide whether or not it is valid, we should take a moment to talk about hydroponics and aquaponics.

Hydroponics

Hydroponics is a method for cultivating plants without soil, using only water and chemical nutrients. The "ponics" in "aquaponics" comes from hydroponics. The term "hydroponics" literally means "water working". Much of the greenhouse tomato, basil and lettuce production in North America today is done using hydroponic growing techniques, but you might have also heard of it because it is the favored growing method of marijuana producers.

Aquaponics is a hydroponic growing method in that it requires no soil. In both methods, the plants' roots are constantly bathed in highly oxygenated, nutrient-rich water, and both see growth rates far above those found in soil-grown plants.

Aquaponics also borrows from many of the classic hydroponic system types. The flood and drain (also known as ebb and flow) style of growing on which this book focuses comes from the hydroponic world, as do NFT (nutrient film technique) and DWC (deep-water culture or raft) styles.

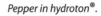
Pepper in hydroton®.

This is where the similarities end, however. Aquaponics is an improvement over hydroponics for the following reasons:

1. Expensive chemical nutrients are replaced by less expensive fish feed. Hydroponic nutrient solutions are expensive, and are gradually becoming more expensive as some ingredients are becoming over-mined and increasingly difficult to acquire. A gallon of hydroponic nutrient solution costs $30–60, and a few tomato plants will easily go through that during their productive lifetime. Meanwhile, a 50-pound (23-kg) bag of tilapia feed costs about the same amount, and at a 1.3 feed conversion ratio will give you 38 pounds (17 kg) of mature tilapia and simultaneously support about eight tomato plants.

2. You never dump out your nutrient solution. Water in hydroponic systems needs to be discharged periodically, as the salts and chemicals build up to levels that become toxic to the plants. This is both inconvenient and problematic, as the disposal location of this waste water needs to be carefully considered. In an aquaponic system, rather than having these problems with chemical imbalance, you achieve a natural nitrogen balance that is the hallmark of an established ecosystem. The water in your system is a critical component that you nurture as part of that balance. In aquaponics, you never replace your water; you only top it up as it

Recirulting aquaculture tanks.

evaporates and transpires (evaporates from the leaves of the plants). This saves both water and time.

3. Maintaining an aquaponics system is significantly easier. I've spent years running both system types, and I can assure you that once cycling (starting the system by building the bacteria base or biofilter) has taken place, an aquaponic system is significantly easier to maintain than a hydroponic system. Hydroponic gardeners are instructed to check the EC (electrical conductivity) with a special meter daily, or at least once every few days. In aquaponics testing, this frequently just isn't necessary. Because an aquaponic system is a natural ecosystem, it will tend to move into a balanced steady state. You will need to check pH and ammonia once a week and the only other check — for the nitrate level — can be run monthly.

4. Aquaponics is more productive. A university study by the Crop Diversification Centre in Alberta, Canada (Savidov, 2005), has shown that after six months, when the aquaponic biofilter is fully established, a grower will see faster and better growing results with aquaponics than with hydroponics.

5. Aquaponics is completely organic. Hydroponics is growing in a sterile, man-made environment. Traditional hydroponic systems rely on the careful application of expensive nutrients made from mixing together a concoction of chemicals, salts and trace elements. In aquaponics, you create a natural ecosystem where you rely on bacteria and composting red worms to convert the ammonia and solid waste from the fish into a complete plant food. It is a necessarily organic process. If pesticides are applied to the plants, the fish will suffer. If growth hormones or antibiotics are given to the fish, the plants will suffer. Aquaponics relies on nature and is rewarded through better growth, less maintenance and lower disease rates.

Aquaculture

The "aqua" in "aquaponics" means "water" and refers to the aquaculture side of the aquaponics equation. The dictionary defines aquaculture as "the cultivation of aquatic animals and plants, especially fish, shellfish and seaweed, in natural or controlled marine or fresh water environments." Clearly aquaponics has a foundation in aquaculture in that the nutrients for the plants come from fish. Many of the early pioneers in aquaponics come from aquaculture academia, such as Dr. James Rakocy, who were initially interested in aquaponics as a way to solve the problem of fish waste disposal. (Bernstein, 2010)

The history of aquaculture actually dates all the way back to the ancient Chinese back in the fifth century BC. They would capture young fish in wild habitats then transfer them to an artificial environment to grow. The Romans were known to have cultivated oysters (are you surprised?) and there are even Egyptian hieroglyphs that are thought to represent intensive fish culturing. (Batis n.d.)

The first known example of "modern" aquaculture occurred in 1733 when a German farmer successfully gathered fish eggs, oversaw their fertilization and hatching, and then raised them to maturity. These techniques were exclusively focused on freshwater fish. Later the practice of creating farming "pens" off ocean shorelines for raising saltwater fish was developed.

The most recent development in aquaculture has been recirculating aquaculture systems or RAS. This is a technique where fish are raised in large, densely stocked tanks. A big advantage of RAS is that it does not require natural bodies of water, so RAS systems can be set up anywhere, even in urban centers. Rather than shipping tilapia or perch thousands of miles across the country, they can now be raised near those who want to cook them for dinner.

Another big advantage is that because of advancements in aquaculture science, fish can be raised very densely in RAS. Stocking densities as high as one pound of fish per gallon of water have been successfully achieved.

On the other hand, RAS is capital intensive, energy intensive and risky. The risk stems from the high packing densities and the derivative need for oxygen-rich water. Aeration depends on systems powered with electricity. Because of the high packing densities, there is little time to act should the power fail. Millions of fish can be killed from lack of oxygen in less than an hour.

The main disadvantage of RAS is the amount of waste the fish produce and, more importantly, the waste disposal process. Fish produce waste through their respiration process, mostly in the form of ammonia, which they excrete through their gills. They also produce solid waste through their digestive process. Another source of waste in an aquaculture operation is the excess, uneaten food that sinks to the bottom of a fish tank. Current filtration methods — be they mechanical, chemical or biological — all rely on extracting the waste from the fish tank and disposing of it as a harmful byproduct. (Wheaton, n.d.)

While aquaponics got its start in aquaculture, it fundamentally departs from the earlier form in a very important way — what is a waste product and a problem in aquaculture is a treasured input in aquaponics.

This is a significant shift in aquaculture philosophy. As the blog post below illustrates, it might be a while before the aquaculture community embraces it. The blog was written by a member of the Aquaponic Gardening Community to recount her experience attending an aquaculture conference as an aquaponic gardener:

Just got back from the 8th annual engineering conference on Aquaculture, Roanoke, VA.

I was impressed by two things:

1) How much they have accomplished in the high density production of fish, and
2) How hard they are working to solve the very problems that aquaponics solves so well. Research papers presented and attended by engineers from 23 countries!

They are investing an enormous amount of time, money and energy (literally, electrical energy) to produce tons of fish. Most of the major issues, DO (dissolved oxygen), waste products and water treatment can be managed effectively with aquaponics but not at the tonnage that they are trying to achieve.

Most of the papers presented were on the order of how to solve those problems within a high-density aquaculture setting; raising fish in isolation.

But why the high density? Why that approach?

The majority of funding, of course, comes from the "industrial" corporations at the university research centers. Think CAFO ... i.e., confined animal feeding operations ... feed lots (cattle, hog, chicken, etc.) that can maximize profits on the smallest footprint.

I saw some stunning results ... tanks full of Atlantic salmon, 8lb each, at least ... in pristine water, with an annual tonnage of 50, being projected, for delivery. Incredibly dependent on very, very high energy input, O_2 injectors (cost of the O_2), fed by the ocean "junk" fish, wheat and corn currently.

When asked what our interest was (my husband & I), many conference attendees were fascinated by the idea of including grow beds to round out a fish growth system! Explaining that we are developing a sustainable farm with grass-fed beef, heritage pork, dairy cows and free range poultry ... and would like to add fish, as well as hydroponic grow beds, in addition to our organic gardens, for our local market.

Our interest is to develop a system that could work in small communities with minimal energy and water use. It did not have to produce tonnage ... just enough for a local food market (i.e., the 100-mile diet). We were repeatedly asked to have information forwarded to them. Many of these requestors where PhDs, MDs and industry experts!

— Amy Crawford, Aquaponic Gardening Community,
blog post entitled "Interesting Challenge —
Moving the Research Community to 'Support' the Local Community"

What is aquaponics? — conclusion

So while a gardener might describe aquaponics as organic hydroponics, an aquarium or pond hobbyist might think of it as an aquaculture system with natural filtration. Both are correct, and both are insufficient. Aquaponics is truly a unique system unto itself. One where nature has stepped in and helped relieve some of the burdens inherent in each system.

In a February 2010 *New York Times* article, journalist Michael Tortorello described Connecticut resident Rob Torcellini's aquaponics setup as "either a glimpse at the future of food growing or a very strange hobby — possibly both." In the next chapter of this introductory section we will talk about how aquaponics fits into the future of food, and the section after that will be about aquaponics as a hobby. Once you read these, perhaps you can then judge for yourself if it is very strange, or simply a unique and relatively undiscovered agricultural treasure.

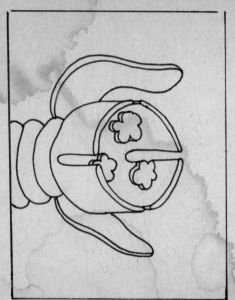

CON AMOR, CK. ♡

2

The global perspective

*"Given the ecological and economic viability of aquaponics, food would
be significantly more just if this unique form of aquaculture
(aquaponics) became the future of floating protein".*

— James McWilliams, Just Food, 2009

This chapter diverges from the rest of this book. In it, I explore aquaponics from a global vantage point instead of focusing on getting you started with your home-based system. You are welcome to skip ahead to the next chapter if you are in a hurry to start your aquaponic garden, or simply aren't interested in how aquaponics can be a part of the solution for our future food supply problems. But I hope you don't. The first part of this chapter is full of gloomy news about trends that are converging to imperil our food supply. I believe, however, that recognizing the bad news is a key first step to becoming an active part of the solution. The second part of this chapter will help you to understand that you are already becoming part of the solution by reading this book and learning about aquaponic gardening.

The bad news

"The most worrying trend in the world today is not terrorism but demographics."

— Michael V. Hayden, Director, Central Intelligence Agency (Friedman, 2009)

I started writing this chapter on New Year's Day, 2011. That day, the headline on the second page of our local paper read "India's Farmers Say Climate

Change Hurting Tea Growers." The climate in Assam state, where 55 percent of India's tea crop is grown, has changed just as it has in the rest of the world. It has become steadily hotter and rainfall patterns have changed, causing an 18.4 percent decrease in tea production since 2007. The temperature in the region has increased 3.6 °F (2 °C) in the past eight decades, and the area has become more cloudy and humid as a consequence. The same article said that the UN science network foresees temperature rising up to 11.5 °F (6.4 °C) by 2100, and that NASA had just reported that the January–November 2010 period was the warmest globally in the 131-year period for which records have been kept. The article ended by pointing out that the United States has long refused to join the rest of the industrialized world and sign the Kyoto Protocol. The reason? Ironically for the tea farmers of Assam, India, we refused to sign because we said it would hurt our economy. (*Boulder Daily Camera,* 2011)

Stories like this not only add to the drumbeat of bad news about how climate change is adversely affecting our food supply, but also make clear that we cannot rely on politicians to solve this problem. Plus, unfortunately, we likely do not have all that much time to find a solution. Many predict that events over the next forty years, largely of humanity's doing, will converge to put into question our ability to feed ourselves.

Future food economics: increased demand

"We must alert and organize the world's people to pressure world leaders to take specific steps to solve the two root causes of our environmental crisis — exploding population growth and wasteful consumption of irreplaceable resources. Over-consumption and overpopulation underlie every environmental problem we face today"
— Jacques Cousteau

Except for Shakespearean studies, my favorite subject in college was economics. I was fascinated that human behavior could be synthesized down to lines on a graph. The first example was the famous graph of the law of supply and demand. It said that if demand rose and supply stayed constant, the price would go up. There were more dollars "chasing" the same number of goods.

Consider the two main drivers of the projected increase in global demand for food in the next forty years: global population growth and increasing standards of living for developing nations.

Population growth

First, demographers project worldwide population will grow from the current seven billion to nine billion by 2050. This means that we will, on average, add 75 million mouths to feed, or the equivalent of the population of Germany, every year for the next forty.

On August 12, 2009, CNN reported that according to the Population Reference Bureau's 2009 World Population Data Sheet, a staggering 97 percent of global growth over the next 40 years will happen in Asia, Africa, Latin America and the Caribbean. Most of the growth in the developed world is expected to take place in the United States and Canada. (CNN, 2009)

Global Population	How long it took to get there
1 billion in 1804	
2 billion in 1927	(123 years later)
3 billion in 1960	(33 years later)
4 billion in 1974	(14 years later)
5 billion in 1987	(12 years later)
6 billion in 1999	(12 years later)
7 billion in 2011	(12 years later)
8 billion in 2025	(14 years later)
9 billion in 2049	(24 years later)

US Census Bureau, International Database

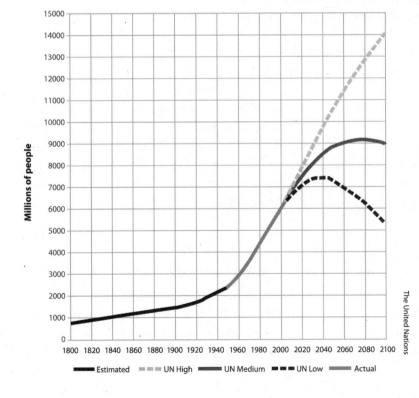

World population from 1800 to 2100, based on UN 2004 projections and US Census Bureau historical data.

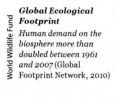

World Wildlife Fund

Global Ecological Footprint

Human demand on the biosphere more than doubled between 1961 and 2007 (Global Footprint Network, 2010)

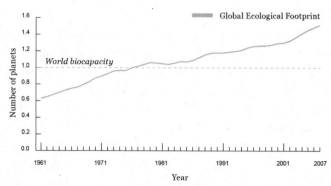

Humanity's ecological footprint.

The same report drew a comparison between Uganda and Canada, which currently have about 34 million and 31 million residents, respectively. By 2050, Canada's population is projected to be 42 million, while Uganda's is expected to soar to 96 million, more than tripling.

So what is the problem with all these statistics? Clearly responding to population growth has been a part of the story of human history. And isn't it good news is that we are growing at a decreasing rate?

The problem is that we are perhaps already beyond our planet's ability to support even the life we have now.

According to the World Wildlife Fund's _Living Planet Report 2008,_ "humanity's demand on the planet's living resources, its Ecological Footprint, now exceeds the planet's regenerative capacity by about 30 per cent" (see the chart above). What this means is that we are now consuming planetary resources faster than they are being regenerated, including the planet's ability to process waste. Now add to the mix that most of the upcoming population growth will take place in areas of the world that are least able to support that growth. Again, from the WWF report: "The resulting deforestation, water shortages, declining biodiversity and climate change are putting the well-being and development of all nations at increasing risk."

Increasing standards of living for the developing nations

The second big impact on world food demand is globalization. Rising standards of living, especially in China and India, will increase demand for an American lifestyle, especially regarding meat consumption. In his 2008 book _Hot, Flat and Crowded,_ Thomas Friedman describes the flattening of the world as the rise of high-consuming middle classes across the planet. The technological revolution that is connecting us all is leveling the playing field and opening up a world of possibilities to everyone, no matter what the development status of their country.

This is opening the eyes of millions across the globe to what a better life looks like and carrying huge change with it. Most of this impact is good. Breaking down geographic boundaries promotes peace because it enables us to better understand and appreciate both our differences and similarities as humans. Sharing information greatly accelerates the pace of innovation. Aquaponics itself is a testimony to this.

But this eye opening is also driving ever-increasing demand. Consumption in the developed world is the primary source of strain on the earth's resources. The problem multiplies when we try to elevate the standards of the developing world to meet those of the developed world. What if the Chinese were able to increase their standard of living to match that of North America? In a *New York Times* essay (January 2, 2007), geographer and historian Jared Diamond pointed out that "What really matters is the total world consumption, the sum of all local consumptions, which is the product of the local population times the local per capita consumption rate. Currently, each of us 300 million Americans consumes as much as 32 Kenyans. Per capita consumption rates in China are about 11 times below ours. If the Chinese consumed the way we do, we would roughly double world consumption rates. If India as well as China were to catch up, world consumption rates would triple. If the whole developing world were suddenly to catch up, world rates would increase eleven fold. It would be as if the world population ballooned to 72 billion people (retaining present consumption rates)."

By any definition of fairness, the citizens of the developing world have every right to improve their circumstances to the point that they are eating from the same magnificent spread that we have been enjoying this past century. But can we really afford to have the entire population at the same consumptive party? Again, some say that we are already beyond the planet's biological capacity to support itself. Imagine if the entire planet suddenly achieved a much higher standard of living.

Future food economics: decreased supply

Now, think back to our simple economic model of supply and demand. We just covered the two reasons why global food demand will increase over the next forty years: population growth and increasing standards of living. Now let's look at why supply — our ability to grow food at the same rate we have been growing it — might decline.

Our current model of industrial agriculture depends on three factors that are no longer in place: cheap fossil fuels, unlimited water and a stable climate. Let's take a look at how each of these inputs affects our food supply.

Petroleum use in agriculture

I saw a very interesting video recently about the owner of a small farm in England who was exploring how she was going to survive in a world of declining energy reserves. The farmer explained how vital fossil fuel is to producing our food by taking apart her sandwich and showing how petroleum was used in each step of its making: (Hoskin, 2009)

- First, you use a tractor to plow and plant the field to grow the wheat for the bread
- Then you add the petroleum-based fertilizer required to grow the wheat crop in depleted soils
- Next you control insects, weeds and disease using petroleum-based pesticides, herbicides and fungicides
- Once the grain is ripe, you harvest it using gas-powered harvesters
- Then you drive the grain to a factory to be dried, which uses even more power and oil
- Then you drive it to a factory bakery to be processed into bread, and then drive it to the store where it can be bought
- The pigs that produced the ham in her sandwich were also energy "hogs." One pig can eat nearly half a ton of the grain that was produced using all the oil discussed above
- The tiny bit of lettuce and tomato were either shipped or flown in or produced in a heated greenhouse; either way they consumed oil to get to their destination

That ordinary sandwich was figuratively dripping with oil.

In her continuing remarks, the farmer brought up the oft-quoted statistic that we use ten calories of energy for every one calorie of food we produce worldwide. This is only sustainable if there is an unlimited supply of cheap, renewable energy. Petroleum, the energy engine of agriculture, is only renewable if you consider time horizons in the millions of years. And while scientists, researchers and activists are debating the exact date of "peak oil," no one disagrees that our supplies are limited. Going back to our basic economic model, if the supply of something is limited and the demand increases, then the price will necessarily rise. Because oil is so intricately woven into every aspect of our current food-production system, increasing oil prices will affect the price of food, perhaps dramatically. It is already happening.

Water use in agriculture

> *"Water, water, everywhere, nor any drop to drink"*
> — from *The Rime of the Ancient Mariner* by Samuel Taylor Coleridge, 1798

Although more than 70 percent of the earth is covered in water, only 2.5 percent of that is fresh water. Of that, nearly 70 percent is frozen in the ice of Greenland and Antarctica. Most of the remainder is either captured in our soil as soil moisture or is in parts of groundwater aquifers too deep for realistic access. Less than one percent of the earth's water is actually available for human use. (University of Michigan, 2006)

And we humans are using that one percent at an increasing rate. In a speech on February 5, 2009, United Nations Deputy Secretary-General Asha-Rose Migiro warned that two-thirds of the world's population will face a lack of water in less than twenty years if current trends in climate change, population growth, rural-to-urban migration and consumption continue. Ms. Migiro also noted that agriculture consumes roughly three-quarters of the world's fresh water and in Africa the proportion is closer to 90 percent. (UN News Center, 2009) In the United States 80 percent of our consumptive water use is for agriculture; the figure is over 90 percent in several Western states. (USDA, 2004)

When the earth had six billion citizens, we used nearly 30 percent of the world's accessible, renewable water supply. Projections for 2025 are that we will be using 70 percent. Just like the rest of our natural resources, water use will not be evenly distributed. (Friedman, 2009)

Freshwater is made available to us in three ways: as rain, surface water (lakes, reservoirs, streams, etc.) and near-surface groundwater aquifers. All three sources are currently being threatened by climate change, overuse and pollution. Ultimately they all flow into the groundwater reserves, which are our water savings account. If the groundwater isn't recharged at the same rate that it is being withdrawn, it becomes depleted and eventually disappears.

About one third of the world is completely dependent upon groundwater, which is the second largest reserve of fresh water in the world, after the polar icecaps and glaciers. As the global population increases, and our need for water follows suit, we will draw down our groundwater supplies exactly as we are drawing down our petroleum supplies. In some areas of China, groundwater levels have dropped at the rate of 1.5 meters per year over the past ten years. (Worm, 2004)

In the United States, 40 percent of our fresh water comes from groundwater supplies. The most famous example is the Ogallala Aquifer that provides water for South Dakota, Nebraska, Colorado, Wyoming, Kansas, Oklahoma, Texas and New Mexico. It spans an area of 800 miles (1300 km) from north to south, and 400 miles (650 km) from east to west. The Ogallala is used mainly for agricultural irrigation in this semi-arid region. Since it was first tapped in 1911, six percent of the aquifer has become unusable because of depletion. At the current rate of draw, scientists estimate that another six percent will become unusable every 25 years henceforth.

Aquifer depletion has drastic consequences that go beyond the obvious lack of water. The land above a depleted aquifer can turn into a "cone of depression" or a sinkhole and become dangerous and unusable. If an aquifer is close to an ocean, lowering the water level can destabilize the barriers between the aquifer and the salt water. If this results in seepage of ocean water into the aquifer, the remaining water in the aquifer becomes unusable. (Worm, 2004)

Worse, the water we can use is being polluted through the very agriculture that it nurtures. In the 2000 National Water Quality Inventory, states reported that agricultural nonpoint source (NPS) pollution is the leading source of water quality impacts on surveyed rivers and lakes, the second largest source of impairments to wetlands and a major contributor to contamination of surveyed estuaries and ground water. What is "nonpoint source" pollution? It is pollution that comes from a diffuse array of sources instead of a single "point," like a factory or a sewage treatment plant. Agricultural activities that cause NPS pollution include poorly located or managed animal feeding operations; overgrazing; plowing too often or at the wrong time; and improper, excessive or poorly timed application of pesticides, irrigation water and fertilizer. Pollutants that result from farming and ranching include sediment, nutrients, pathogens, pesticides, metals and salts. The consequence is widespread water pollution and degradation of our lakes, streams and groundwater. (US Environmental Protection Agency, 2005)

Climate change and agriculture

"If we want things to stay as they are, things will have to change."

— Thomas Friedman, *Hot, Flat and Crowded,* 2008

A joint statement by 21 national science academies to the 2007 G8 summit declared, "It is unequivocal that the climate is changing, and it is very likely that this is predominantly caused by the increasing human interference with the atmosphere." (National Science Academy, 2007) Our activity on this planet is causing the earth to become, on average, warmer. The eleven hottest years since thermometer records became available in 1860 have all occurred between 1995 and 2011, with 2010 being the hottest of them all.

This change in our climate threatens our ability to produce food around the world. The Food and Agriculture Organization (FAO) warns that a global temperature increase of two to four degrees Celsius over preindustrial levels could reduce crops yields 15 to 35 percent in Africa and Asia and 25 to 35 percent across the Middle East. (Smith, 2010)

But climate change is more than a warming trend (which is why the term "global warming" is an inaccurate description of the phenomenon). Increasing temperatures will lead to changes in many aspects of weather, including wind patterns, the amount and type of precipitation and the types and frequency of severe weather events. Such climate destabilization could have far-reaching and/or unpredictable environmental, social and economic consequences.

Surprisingly, agriculture is probably the single biggest contributor to climate change. When we think of the harvest from agriculture we think of grains, vegetables and beef but with current industrial agricultural methods we are also harvesting carbon dioxide (CO_2), nitrous oxide (N_2O) and methane (CH_4). Carbon dioxide is released by all the fossil fuels described earlier in this chapter. Nitrous oxide comes from chemical-based fertilizers, and has 296 times the global warming potential of CO_2. Methane comes from livestock operations and has up to 25 times the global warming potential of CO_2. (Smith, 2010)

Deforestation

More than 40 percent of the earth's land has already been cleared for agriculture. Agriculture currently uses 60 times more land than urban and suburban areas combined, and covers an area of the earth the size of Africa. Yet it probably isn't enough to feed the growing demand for farmland. In our quest for more and more fertile soil, we are turning to the soils that have been created on the floors of the tropical rain forests.

According to Rainforest Action Network, more than an acre and a half of tropical rain forest is being cleared every second of every day. This means that every week, we lose an area the size of Rhode Island, and every year an area twice as big as Florida is destroyed. At this rate the rain forests will be entirely gone by 2060. (Save the Rainforest, 2005)

Ironically, by clearing tropical rain forests to create more farmland to feed ourselves, we are tearing down our best chance to solve the climate problem and filter our air. The rain forests are often described as the "earth's lungs" because of their tremendous efficiency in converting carbon dioxide into oxygen. They are the single greatest source of the air we breathe.

Overfishing the oceans

Before I move on to aquaponics as part of the solution, I need to raise one more critical food supply issue: the overfishing of our oceans.

Our oceans are arguably the last wild source of food on our planet, and we are quickly emptying them of fish (*The End of the Line*, 2009):

- We have lost 2,048 fish species that we know of due to overfishing.
- In March 2009, the Food and Agriculture Organization of the United Nations reported that more than 70 percent of fish species were currently endangered.
- In a study at the National Center of Ecological Analysis and Synthesis at the University of California, scientists projected that, barring significant changes, the oceans would become barren of fish by 2048.
- According to some estimates, 85 to 95 percent of the fish caught by commercial fishers is bycatch (aquatic life accidentally harvested by trawlers).

We only end up eating about 10 percent of all the marine life that is killed in order to feed us. (Backyard Aquaponics magazine, 2009)

Animal	Food conversion ratio, lb food/lbs growth
Fish	1.5 : 1 (McGinty n.d.)
Poultry	2 : 1 (Hamre, 2008)
Pigs	4 : 1 (Losinger, 1995)
Cattle	7 : 1 (Cal Ag Ed, 1990)
Sheep	8 : 1 (Cal Ag Ed, 1990)

Feed conversion ratios.

The depletion we are seeing is happening because we have become so incredibly efficient at harvesting throughout the entire depth of the ocean that is populated by fish we eat. According to the 2009 film *The End of the Line,* we now have the technology and fishing fleet capacity to catch four times the current supply of fish. We have enough long-line hooks to encircle the globe 550 times. You could fit 13 jumbo jets through the mouth of the largest trawling net.

To solve this crisis, various governmental agencies have attempted to set annual quotas by species to prevent a complete collapse. For example, the International Commission for the Conservation of Atlantic Tuna (ICCAT) sets the limits on bluefin tuna. According to the producers of *The End of the Line,* however, these limits are set far above recommended recovery levels, and are largely ignored in any case. It is estimated that at least 50 percent of the fish we eat from the ocean was caught illegally.

So, we need to either stop eating so much fish or turn to aquaculture and farm more of our fish on land so that the ocean species can recover.

It's doubtful that we are going to eat less fish anytime soon. The health benefits of fish are just too compelling. Eating fish can provide an excellent source of omega-3 fatty acids, vitamins and minerals that benefit one's overall health. The American Heart Association recommends at least two servings of fish per week to help prevent heart disease, lower blood pressure and reduce the risk of heart attacks and strokes. Compared with other sources of animal-based proteins, all of which are full of saturated fats, fish is the healthy alternative.

Fish are also vastly more efficient sources of protein than other forms of animal protein. Currently 37 percent of the world grain harvest is being used to produce animal protein. (Brown, 2003) Unfortunately most of that is going to feed the most inefficient source of animal protein: cattle. Look at the food conversion chart above and imagine what would happen if the 700 million tons of grain used annually for livestock feed were fed to fish and chicken, rather than pork and beef.

Why is this so? In his 2009 book *Just Food* James McWilliams explains it eloquently:

> The ability to float effortlessly negates the need to undertake such energy-hogging endeavors as standing, walking and running. All these forms of mobility are internal energy sinks for terrestrial creatures, but not for fish. A fish is ecologically honed to translate the majority of its caloric intake into ample flesh that's edible, usually tasty, rich in protein and flush with heart-healthy oil. This streamlined translation, if responsibly managed and harnessed by humans, has the potential to improve the environment while generating more protein with fewer resources. (p. 155)

The global demand for fish has increased dramatically. In its 2010 report on "The State of the World's Fisheries and Aquaculture" the FAO revealed that global average consumption of fish has hit a record of 37 pounds (17 kg) per person per year. And for the first time, nearly half of that fish is being supplied through aquaculture. "Fish farming is the fastest growing area of animal food production, having increased at a 6.6 percent annual rate from 1970 to 2008," the agency said in the report. "Over that period, the global per-capita supply of farm-raised fish soared to 17.2 pounds from 1.5 pounds." (Jolly, 2011)

Unfortunately, while aquaculture clearly relieves some of the strain on the supply of ocean fish, it doesn't relieve it all. The feed used in aquaculture operations uses ocean-harvested fish to create the fish meal that is its protein base. While there are several universities and private organizations trying to find viable alternative sources, as of this writing, none were currently available commercially.

In addition, fish farming, whether done in cages and pens floating off the coastline or in recirculating tanks inland, produces substantial amounts of waste. These farms have even been described by some as "floating CAFOs" (confined animal feeding operations), comparing their waste production and pollution with that of the cattle and poultry industries.

Finally, while you would think that the ability to grow fish in ponds and tanks would be a boon for local fish production, in fact, here in the US only 10 percent of the farmed fish we eat is produced domestically. China produces 62 percent of the farm-raised fish in the world today. (Jolly, 2011)

The bad news summary

As I finish writing this chapter, I am reflecting on another news article that was published yesterday on MSNBC. It is entitled "Global Food Chain Stretched to the Limit." It describes a food system on the brink of collapse because of rising demand, rising oil prices and climate change. The FAO reported that its food price index jumped 32 percent in the second half of 2010, soaring past the previous record set in 2008. The president of Tunisia was chased from his country by rioters demanding lower food prices. "Situations have changed. The supply/demand structures have changed," said Abdolreza Abbassian, Chief Economist at the FAO. "Certainly the kind of weather developments we have seen makes us worry a little bit more that it may last much, much longer. Are we prepared for it? Really this is the question." (Schoen, 2011)

The good news

"Though I travel around the world and I see environmental degradation and horrors ... I'm filled with hope when I come to a place like this."

— Philippe Cousteau, October 22, 2010, while touring the aquaponics system
at Sickles High School in Tampa Bay, Florida

Enough of the doom and gloom. Now let's talk about some good news and how to solve, or at least work around, the problems I outlined above.

While aquaponics can't address the dual demand pressures of population growth and increasing global standards of living, it does offer many exciting solutions to many of the production problems we face.

Petroleum use in aquaponics vs. traditional agriculture

While some energy is needed to heat the water and power the pumps that move the water and supply oxygen to the fish, that energy can generally come from renewable sources. Aquaponic gardeners are already using sustainable sources of heat like geothermal, solar and rocket heaters to heat their fish tanks. They have also discovered that techniques including insulating, burying and covering their fish tanks go a long way in preventing heat loss.

Because there is no soil to till, there is no longer a need to use tractors and gas-powered farm equipment. Commercial aquaponics operations typically employ either a raft method, where the plants float in water until they are harvested, or media. Neither requires the kind of labor that soil-based farming does. Since there are no weeds in aquaponics, there is no need to mechanically remove weeds or spray herbicides. Since the plant nutrients and water are both integral to an aquaponics system, there is no need for petroleum-based fertilizers or truck-mounted irrigators. Since aquaponically grown plants are either growing in waist-high grow beds or in rafts floating in water, they are much easier to harvest than soil-grown plants.

Aquaponic farms will also be important in a reduced fuel world because they are completely site agnostic. Aquaponic systems can be set up anywhere you have, or can artificially establish, an appropriate climate for the plants. Poor soil? No problem. Aquaponics is particularly well adapted to providing food to local communities that might not otherwise have fertile land available for growing. Since over half of humanity now lives in our cities, it is important that food-growing facilities be established where the people are, rather than trucking food in from distant locations. Currently, most of our produce is shipped hundreds, if not thousands of miles. Imagine how much fuel could be saved if we actually grew our food in our city centers.

Water use in aquaponics vs. traditional agriculture

This is where an aquaponics system really shines.

Modern agricultural methods waste an incredible amount of water. Water is either sprayed or flooded through fields where a huge amount either evaporates into the air on a hot day, or seeps past the plant roots and into the water table, pulling chemical fertilizers, herbicides and pesticides down with it.

Aquaponics, on the other hand, is a closed, recirculating system. The only water that leaves the system is the small amounts taken up by the plants (some of which transpires through the leaves) or that evaporates from the top of the tank. That's it. You can see why it is easy to believe that aquaponics uses less than a tenth the amount of water a comparable soil-based garden uses.

Aquaponics is even more water thrifty than it's horticultural cousin, hydroponics. Because hydroponics is a completely human-managed, chemical-based system, the nutrients regularly become unbalanced. The nutrients that the plants aren't using at any particular stage in their development build up to toxic levels. Every two to four weeks, the entire nutrient solution reservoir needs to be pumped out and replaced with fresh chemicals. The nutrient waste from hydroponic systems is full of chemical mineral salts that need to be carefully disposed of and prevented from running off into our streams and seeping into our groundwater.

Since aquaponics is an organic ecosystem in which the nutrients are balanced naturally, there is never any toxic buildup of nutrients. In fact, because the water in an aquaponics system is so full of healthy biology, I recommend that if possible, you never discharge the water from your fish tank. The only reason why you ever would is if something caused extreme amounts of ammonia to overwhelm your biofilter's ability to convert it and you therefore needed to do a partial water change to dilute the ammonia. An example of this would be a dead, decomposing fish that you were unaware of. Even if such a rare event were to occur, the discharge from your aquaponics system is completely organic and will only benefit any soil lucky enough to be watered by it.

Climate change and aquaponics vs. traditional agriculture

I won't claim that aquaponics can scrub CO_2 from the atmosphere or help restore stability to our climate. I am very comfortable asserting, however, that an aquaponics system is a food-growing system that could have zero impact on our environment, especially if the pumps and heaters are powered through renewable energy sources. Except for purely wild food-growing systems, such as the ocean, and most permaculture techniques, no other food system that I know of can make that claim.

On the other hand and as discussed above, traditional agriculture is the single largest contributor of CO_2 emissions, while simultaneously contributing to the ongoing shrinking of the earth's CO_2 filter through the need for more and more land for growing crops and raising cattle. The main pollutant sources are CO_2 emissions from all the petroleum being used in farm production and food transportation, methane from cattle production, and nitrous oxide from over-fertilizing. Aquaponics requires none of these inputs. Petroleum needs in aquaponics range from much less to zero. Fish don't produce methane as cattle do, and there is no chance of over-fertilizing an aquaponics system.

Perhaps most importantly, aquaponic systems can be started anywhere. So now instead of clearing jungles and forests we can instead focus on our urban centers and begin to think of old factory and warehouse buildings as the farms of our future. While perhaps not suited to growing vast fields of grain, aquaponics can now grow any vegetable and many types of fruit crops, and do it in a way that is even more productive on a square foot basis, even in an urban setting. Aquaponics can produce 50,000 pounds of tilapia and 100,000 pounds of vegetables per year in a single acre of space. By contrast, one grass-fed cow requires eight acres of grassland. Another way of looking at it is that over the course of a year, aquaponics will generate about 35,000 pounds of edible flesh per acre, while the grass-fed beef will generate about 75 pounds in the same space. (McWilliams, 2009)

Is the notion of producing at least some portion of our food in our urban centers a science fiction fantasy? Not at all. In fact, in her essay "Reconsidering Cities," author Sharon Astyk pointed out that it isn't as unusual as you might think for city dwellers to grow a meaningful portion of the food they eat. She explains that Hong Kong and Singapore already both produce more than 20 percent of their meat and vegetables within the city limits. In 2002, with more than six million people, Hong Kong was producing 33 percent of the produce, 14 percent of the pigs, 36 percent of the chickens and 20 percent of the farmed fish eaten in the city limits. Will cities eventually grow all their own food? No, but they don't necessarily have to. A substantial portion can be enough, as long as they also build strong ties to surrounding rural areas. (Astyk, n.d.)

Aquaponics vs. aquaculture

With the projected dramatic decrease of the bounty of the oceans, we must turn to aquaculture if we are to continue to enjoy the healthful benefits of eating fish. The central problem is, as with any intensive animal-raising operation, how to get rid of the waste without harming the environment. Again, aquaponics by its very definition solves this problem. Aquaponics takes the potentially toxic waste water from an aquaculture system and creates an organic nutrient for a hydroponic system. This acts as a biofilter for the aquaculture system and purifies the water that goes back to the fish. By seeking a solution through bio-mimicry techniques and observing nature, scientists found the solution in polyculture. By intertwining the fish culture and the plant culture through a man-made recreation of natural wetlands systems, aquaponics was born. In an aquaponics system, nature plays the lead role and the successful farmer conducts rather than dictates.

The global perspective — conclusion

*"When the wind changes direction, there
are those who build walls
and those who build windmills."*

— Chinese proverb

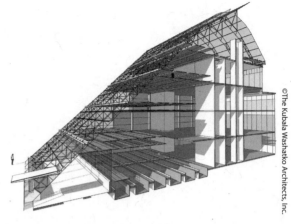

©The Kubala Washatko Architects, Inc.

Growing Power vertical farm.

On the day I finished this chapter, I read a newly published story on urban agriculture, and this one filled me with hope. Mayor Barrett of Milwaukee just gave the final approval for a zoning variance to enable Growing Power to build the world's first vertical farm. The five-story structure will be built to grow plants and fish together, with the water flowing from one story to another in a vision that reminds one of the Hanging Gardens of Babylon. It is a building entirely dedicated to the growing of fresh food in an area of Milwaukee surrounded by low-income housing, liquor stores and mini-marts.

The Chinese proverb above provocatively asks each of us a question. As the winds change, and the repercussions of past sins of humanity's resource over-indulgence become more and more a part of our everyday lives, will we build walls or windmills? By reading this book and learning more about aquaponics you are choosing a windmill. You are choosing to become a part of the solution.

3

Home food production

*"Aquaponics is a 'way of life' that enables growing food
for oneself and the community, without the use of
harmful pesticides and herbicides, while using
the least amount of resources thereby leaving
the smallest carbon footprint."*

— Sahib Punjabi, Lake Mary, Florida

So aquaponics from a world perspective is all very interesting, and yes, aquaponics has the potential to be an important part of our future global food producing system. But this book is about home gardening and producing food at the individual and family level. Are there any advantages to aquaponics for someone who simply wants to take this up as a hobby and/or to help feed their family? Yes!

You are probably considering whether or not to take on aquaponics as a new adventure, but might need to convince your spouse or others close to you that it is going to be worth the time and money. Or perhaps you are already fully committed to an aquaponics project and want to convince your friends and neighbors to start their own systems. Or perhaps you are planning on trying to convince your child's school or the local senior center to build a greenhouse and need material for a presentation. What follows is an unabashed list of all the reasons why anyone (and everyone) should have an aquaponics system.

Earth-smart gardening

Since we just finished with a chapter that focused on the environment, I thought it was most fitting to start our list with a reminder of some of the reasons why aquaponics is an earth-smart way to garden. In his book *Agenda for a Sustainable America,* John Dernbach wrote "something is environmentally or ecologically sustainable when it protects, restores, or regenerates the environment rather than degrades it." I believe the list below upholds those simple values.

- Necessarily organic produce — If you use pesticides or herbicides, you will harm your fish and bacteria. If you use hormones or fish medicines you could harm your plants. An aquaponic system necessarily produces food that is free of chemicals. It can't work any other way.
- Grow using one-tenth the water of dirt gardening — Because the water in an aquaponic system recirculates rather than seeping into the groundwater, aquaponics uses far less water than traditional soil-based gardening. It is also far more water thrifty than hydroponics because the nutrient solution is never dumped and replaced.
- Turning a waste disposal problem into a valuable input — Aquaculture treats the waste the fish produce as a harmful byproduct to be disposed of. Aquaponics turns that around and treats the waste as a valuable input into the plant growing part of the system. In nature there is no waste.
- Growing plants hydroponically without hydroponic chemical fertilizers — Aquaponics offers the benefits of hydroponics — fast, closely spaced growth in a dirt- and weed-free environment — without the need for discharging chemically saturated nutrient solution on a regular basis.
- Growing your own food — The average distance most produce in the U.S. travels is 1,800 miles (2,900 km). When your food comes out of your backyard no fossil fuel is used to transport it.

Convenient gardening

In many ways aquaponics is a much easier and more convenient way to grow plants than the traditional gardening you probably grew up with.

- Waist-high, so there is no bending or digging involved — This is actually one of my favorite characteristics of aquaponics. You can set your grow beds at whatever height suits you, and your back. Let's face it. We are all getting older and I for one find that as I age, the appeal of crawling around and turning soil no longer extends much beyond the first warm days of spring. Now with aquaponics, I'm no longer kneeling in the dirt to tend my plants. I merely stroll amongst my grow beds, completely spoiled.

- This is also a terrific benefit to anyone in a wheelchair or who is otherwise physically challenged. Aquaponic systems can easily be made wheelchair-compatible.

- Waist-high, free from deer, dogs and bunnies — Another nice benefit of the grow beds being elevated is stopping pesky herbivores from getting the pick of your garden before you do. No matter where I've lived I've always had to battle with deer for my vegetables, and most years they've worn me down by the end of the summer. Now my grow beds are either in a greenhouse or on a second-story deck where not even the most audacious buck would dare to enter.

- Weed-free — The other gardening chore that regularly wears me down by the end of the summer is weeding. Another reason why aquaponic gardening is so blessedly wonderful is that there are no weeds. I repeat: You will not need to pull weeds from your aquaponic grow beds.

Soil gardening can be tough on the back.

- No dirt — I'm not a soil scrooge. Really. I have as much respect for the complex living organism that is soil as anybody. But a downside of dirt is that it is, well, dirty. It gets on your hands, your clothes and your produce. Aquaponics is a much cleaner way to grow.

- No watering — This may seem self-evident, but is worth mentioning here as an important benefit of aquaponics. Autopsies on most deceased plants reveal that the cause of death was either under- or over-watering. Because of the automatic nature of a recirculating aquaponics system, this risk is now permanently removed from your gardening life. You will never over- or under-water your plants again.

- No fertilizing — Under- and over-fertilizing is another leading cause of plant death and poor performance. Again, because of the nature of the system, fertilization happens automatically and is no longer a gardening chore or risk.

- Can be located anywhere there is sunlight — A traditional, soil-based garden can only work if the garden space you want to grow in is in an area that receives sufficient sunlight for the plants you want to grow. In North America or Europe, it should be on the south or east side of your house, located on level ground that is relatively free of rocks and other debris. And no trees, large shrubs or other obstructions can block the sunlight. But what if you don't have a match between your best dirt and the sunlight? With aquaponic gardening, you can place your garden wherever the optimal sunlight conditions are. You can separate your garden from the ground and place it wherever it will grow best.

Year-round gardening

Because aquaponic gardens can be grown anywhere there is light, whether natural or artificial, they are particularly well suited for indoor and greenhouse growing.

- Larger systems can be transported into a garage or basement for the winter — While it would definitely be an afternoon project, an aquaponics system is portable. If you live in a climate that is unsuitable for year-round outdoor growing, you can design your system from the beginning to be moved indoors for the winter. Smaller grow beds, perhaps on rolling carts, can be detached and relocated. The water from the fish tank can be pumped through a hose into either a second, indoor fish tank or a temporary storage area so the original fish tank can be moved.

- Smaller systems can go anywhere indoors — For those of you not blessed with a suitable outdoor space for growing, you can now easily grow indoors with an aquarium-based aquaponic system. Small, decorative and portable, these systems are perfect for gardening in urban apartments, classrooms, fire stations, even the North Pole!

- Ideal for greenhouses — If you are fortunate enough to have a greenhouse on your property, or are considering building one, aquaponics is quite possibly the perfect greenhouse growing method. Most modern, sustainable greenhouse designs include one or more large containers of water to act as a heat sink, absorbing heat during the day and giving it off at night. With an aquaponic greenhouse, this heat sink is the fish tank, which also becomes the automatic watering and fertilizing systems. Plus, the fish give off CO_2 and their tanks create humidity, so you don't need to supplement either of those.

Growing fish for food

Then there is the fish side of the aquaponics equation, and all the benefits that come with growing your own edible fish.

Basement aquaponics.

- Convenient — Just as a vegetable gardener becomes used to having his produce grown close to his kitchen door, an aquaponic gardener enjoys both this convenience and the convenience of the protein portion of his meal at his fingertips as well.
- Fresh —.Besides being convenient, the at-home accessibility of fish also means that it is as fresh as it can possibly be. An aquaponic fish can literally go from swimming in the water to sizzling on your dinner plate in twenty minutes.
- Safe — In addition to being convenient and fresh, an aquaponically grown fish is safer to eat than any other fish you can buy or catch. You know what this fish has eaten. You know what its growing and harvesting conditions were. You know when it was harvested and how it was stored (if at all). You now have complete control over every factor important in the freshness and safety of the fish you eat. Furthermore, because fish are cold-blooded animals they are not compatible with *E. coli* or *Salmonella* bacteria, so these two horrible food safety concerns are not a part of an aquaponic gardener's set of worries.

Ecofilms Australia

Barbeque fish dinner.

- Food independence — All of this adds up to a big step down the path to food independence. You are freeing your food sources, both produce and protein, from the petroleum-intensive distribution methods that currently get most of your food on your table.
- Fun — Raising fish is just plain fun. A large fish tank churning with hungry fish will add an indescribable element of entertainment and life to your backyard or basement.
- Ecological — The best part is that even with all these benefits you are not asked to compromise any environmental principles. You are lessening the demand for fish from our oceans. You are not patronizing wasteful, polluting aquaculture farms. You are not using energy to ship frozen fish from faraway lands.

No matter what the motivation that brings you to aquaponics, just be forewarned that it is a thoroughly engaging, captivating endeavor that could change your life. At this point it's not too late to turn back. Still want to grow your own vegetables and fish together? OK, let's get started…

Section 2

The plan

4

Before you start

*"Aquaponics is solar-powered nanotechnology that produces fresh
vegetables and meat, while purifying water. This makes it at once the
coolest, most mind-expanding and most IMPORTANT activity
I have undertaken in the half-century I have been alive ..."*

— Rick Op, Houston, Texas

Since you have gotten this far, I am going to assume that you are interested
in creating your own aquaponics system. That's terrific! The rest of this
book will help you reach that goal.

But before gathering tanks, pumps and fish, take a moment and consider
the four questions below. They will help you to gain some clarity on the
"Why," "Where," "When" and "Who" of your aquaponic garden. Having
thoughtful answers to these questions will lead you to create a system that
will serve your needs for many years.

Why are you doing this? Are you a tinkerer or a gardening guru? Is this
something that looks interesting to experiment with? Are you doing it for a
classroom or home schooling environment to show children an alternative
way to grow fish and plants? Do you aspire to reduce your grocery-shopping
bill and provide a significant amount of your family's fresh produce and
fish? Are you considering becoming a commercial aquaponic farmer, but you
(sanely) want to start out small to learn the basics first? Keep the answer to
this "why" question in mind as you go through the rest of the book because

you will find that it will be key to informing all manner of decisions in designing your system.

Where will your system be located? We go over how to think about your system location in some detail in the next chapter, but you should start pondering a few key questions now. Do you have a year-round growing climate? Do you have access to a greenhouse? Can you build a greenhouse on your property? How would your spouse feel about a garden in the basement or garage? Are you growing in a basement with a low ceiling? Will your system be set up on a deck or porch and need to be moved indoors for the winter? Do you live in a high-rise apartment with weight restrictions and aesthetic concerns? The size, aesthetic and location requirements and/or limitations of your aquaponic garden are going to dictate much of your system design. Because the bacteria that are the real engine of the system take a while to develop, and will be killed off by freezing temperatures, you will ideally want to run your aquaponics system all year round. Where can you do that?

When do you want to get started? Again, the bacteria take a while to establish, and that time period is dramatically accelerated by warm temperatures. You will also find that many hatcheries won't ship fish during the winter months. Certain plants can be difficult to establish during short, cool months. Take all the living elements (the "software") into account as you plan for your aquaponics system.

Who will be taking care of this system? Are they particularly tall or short? You can adjust your grow bed heights accordingly. Are there small children involved? You might want to build a window into your fish tank or find one that already has a window. Is there someone, or several people, in wheelchairs? Again, the grow beds can be sized and the height adjusted for maximum accessibility. Does the caretaker travel often? If so, you will want to install backup aeration systems and design for maximum robustness and minimum maintenance.

The plan

The next two chapters in this section will help you to decide where to locate your aquaponics system and select a design configuration for your system based on that location and the other answers you gave above. We will also talk about grow lights, which are an important part of most indoor growing environments. If you have decided to grow exclusively outdoors, or in a greenhouse without supplemental lighting, you can skip the "Lights" section.

Then the rest of this book will address the "How" and the "What" of your aquaponics system. To do this, it is organized into three sections: the hardware, the software and the system. Before we dive in, let's talk about how these next three sections are laid out, and why they are in the particular order they are in.

The hardware

I think about all the apparatus of an aquaponic garden, including the water, as the hardware, because together these elements provide the structure, or the home, for the living parts of the eventual system. Think of them as you would the hardware components in your computer — the monitor, the CPU and the keyboard all provide the necessary structure for the software that brings your computer to life. The hardware side of any system must be built before the software, or living parts, can be established and _____ ely dictate the type of software that can be "run" on the system.

Given this, we will start with a discussion about gr_ This chapter will ask more questions about your system. How much do you want to grow, an to do it in? Are you growing fish for foo_ not? It will then help you select an then help you to find sources f

The next chapter is abo_ or arteries that will connect th will start by talking about the he into detail about the various option fish tank, the grow bed and back to t

After connecting the grow bed an will be time to fill the grow bed with me The next two chapters will guide you throu considerations.

If, in your mind's eye, you were building your _d-ing, at this point you would have a fully operational a _n. The grow bed and fish tank would be connected through plu_ _here would be planting media in the grow bed ready to receive plants _d dechlorinated water in your fish tank. You would be ready to focus on bringing your system to life, so that is where we will go next.

The software

I named this section "software" because just as the software in a computer gives it a raison d'être, so too do the plants, fish, bacteria and worms bring an aquaponics system to life. The fish start the process by introducing the waste that attracts the bacteria and becomes the food for the plants and the worms. The bacteria join the chorus and unite with the worms to release the nutrients the plants seek. The plants act as the cleanup crew for the water that returns to the fish, and the cycle continues.

The chapters in this section each describe a critical living component of your aquaponics system and the role it plays. The Fish chapter explains how to select the fish for your system, where to find them and how to care for them. The Bacteria chapter talks about the kinds of bacteria that will naturally be attracted to your system and how to invite them in and keep them happy. The Plants chapter compares aquaponic gardening to more familiar dirt gardening, then guides you through selecting and starting plants and how to care for them.

Tilapia.

The Aquaponic Source, Inc.

I talk about each one of the main living elements in these chapters without actually encouraging you to go out and obtain them. Not yet. The introduction of each of these elements is potentially tricky and should not be done haphazardly. For that, you need to read the next section.

The integrated system

This is when you "flip on the switch" and your aquaponic system starts its engine! The first chapter, on Cycling, talks you through the process of establishing the bacteria that make up the biofilter in your system both using fish or, alternatively, without fish. You will also learn when to add the plants and when to add the fish if you are cycling without fish.

Tawnya and JD's aquaponics story

We started aquaponics as a fluke. Our mission was to feed our family as much as possible from our backyard. A garden, chickens, even a milk goat were already in the plans for our little homestead, and then we joked that maybe we could turn the kiddy pool into a fish tank. This was a dinner conversation. By midnight we had discovered aquaponics. Entering the cyber world of aquaponics was our first little foray into the possibilities that growing fish and plants together could provide. We spent the next week of sleepless nights and stolen moments around all the other demands of the day, Googling anything and everything we could read. Quickly we came to realize what an amazing concept this was, not to mention the growing number of enthusiasts in search of the very same things: good quality food, no pesticides, synthetic fertilizers or chemicals, naturally produced diverse food products, easy to plant and maintain, better for the planet, really super cool and interesting. The possibilities seemed endless.

The same week we purchased 20 aquariums, hydroponic grow beds and a grow light off Craigslist, along with dozens of trips to Lowe's for plumbing parts. We built a system, made mistakes, redesigned and built some more. We put in the fish (some died, some lived), planted some seeds (some died, some lived) and watched in amazement as food started to grow. When the salads started rolling in, the excitement started to mount. Oh this was going to be great. But one system just wasn't going to be enough. Our sleep was seriously being altered by imagining the possibilities of various designs.

That same month we saw a *Denver Post* article for GrowHaus, a greenhouse in Denver that was going to be renovated for food production using aquaponics. At that moment, we knew this went beyond our basement. Feeling that this was a profession worth pursuing, we read every book, watched every video, and attended every workshop to glean information from the pioneers in the field. With the help of volunteers, we started building bigger systems at the GrowHaus and connecting with a community in a "food desert."

Now our mission is beyond just feeding our family, it is about feeding lots of people. Everyone deserves quality food. Installing these systems in schools and communities, training others, creating job opportunities, showing people the benefits and possibilities of aquaponics, which is what makes every day a great reason to keep feeding the fish. Believing that aquaponics has the potential to go beyond the backyard and become a viable and sustainable way to produce food in a city or out in the country, where water is scare, or the soil is unsuitable, or just because it is an amazing example of mimicking nature: now that is truly inspirational. Share your aquaponics passion with kids, with the elderly, with your neighbors, with a community in need, with everyone. Sustainable food production is cool and we are all a part of it together. FIN!

— JD and Tawnya Sawyer, Lakewood, Colorado

The next chapter is about what you need to consider to maintain a healthy system for the long run. How often should you check pH and clean the pipes? Do you ever need to check nitrates? All that will be covered here.

The main part of the book concludes with a summary of the Aquaponic Gardening Rules of Thumb and, in the Troubleshooting chapter, guidance on what to do if things go wrong. There are some fun and interesting articles in the appendices as well, but once you have read the chapters in the three sections described above, you will have all the tools you need to create, start and successfully run your own aquaponic garden.

5

System location and environment

> *"In our society where even the most essential jobs and products are treated with an enterprising eye, corporate farms are simply depleting our foods of nutrition. The soils are not being replenished and the land is dying. Aquaponics changes how all that works. No pesticides are used and water is retained and reused. Aquaponics is a sustainable, ecologically friendly means of producing nutritious fruits and vegetables as well as filling in another depleting reserve: our oceans' fish."*
>
> — Daniel E Murphy, Middleburg, Florida

This chapter is about the conditions that surround your aquaponics system rather than the environment within the system itself. Because your plants and bacteria are affected by the surrounding conditions more than your fish (you create every aspect of the fish environment whether they are outdoors or in) this chapter is mostly about the seasonal factors that affect plant and bacteria growth, such as air temperature and light availability. How much control you have over these factors depends entirely on the climate where you live and where you decide to locate your aquaponics system.

Climate considerations

I'm envious of the guys in Australia who developed the first home or "backyard" aquaponic systems. After Australian Murray Hallam, the star of the

Aquaponics Secrets and *Aquaponics Made Easy* videos, spent four days with me in Boulder he wrote in his blog:

I have to admit, I had no idea as to the difficulty of year-round gardening in climates where there are severe winters and very definite seasonality (compared to our Australian climate). Only our most southern island state of Tasmania would have similar winter temperatures and climate, as is experienced in much of central and northern USA. Here in Southern Queensland, we can grow all year round, and we can run our aquaponics systems outdoors if we wish. I must add that better results are obtained when the system is in a greenhouse of some sort. It protects the system from heavy downpours and assists in pest and bug control.

With their largely year-round climate, the term "backyard aquaponics" makes sense as a way to differentiate it from "commercial aquaponics." Here in North America, however, that would be far too limiting. Only in Hawaii, southern California and parts of Texas and Florida do you really have a chance of running an aquaponics system year-round outside in a backyard. Even most of Arizona is too hot to practically run an aquaponics system in the summer, unless the fish tank is in a shaded area or inside altogether.

For the rest of us — where freezing temperatures are the norm throughout the winter, or extreme heat the norm throughout the summer (like those in Phoenix, Arizona, for example) — we have four choices:

1. Harvest our plants, harvest our fish and shut everything down for the winter (or summer)
2. Grow indoors
3. Grow outdoors in the summer and bring our aquaponics system indoors for the winter (or summer)
4. Erect a greenhouse

Harvest our plants, harvest our fish and shut everything down for the winter (or summer)

This is a pretty ugly scenario for an aquaponic gardener. First, you need to start as early in the spring (or fall) as possible in order to get a maximum jump on the growing season. But you need to be sure that the temperature is consistently above 65 °F (18 °C) and that the danger of a killing frost has passed or your system will never cycle. Hopefully you will have planned far in advance and will not start your system on a random weekend whim, like I did with my first system.

So let's say you can start your system in Colorado in early May and it takes six weeks to "cycle" (start the biofilter), and you are using the fishless cycling technique we discuss in a later chapter. You can now introduce your fish, which you have hopefully selected to match the temperatures you will be experiencing over the summer. In Colorado, a good choice is tilapia. The most readily available tilapia will be fingerling size (about two inches long). Those will take 9 to 12 months to grow out to harvestable, plate-size fish, depending on their original size, how ideal their conditions are and their feeding regime. This takes you into sometime between March and June of the following year! Your plants will have long since died, and the power bill for heating your fish tank over the winter will be extreme!

You will have the same kind of issue if you decide to use a colder-water fish like trout in the fall. Even if you can keep the trout alive with a pond heater, the plants and bacteria will die off if the temperatures go below freezing.

And then there are the bacteria. In Chapter 13: Bacteria and Worms I go into great detail about the secret lives of bacteria, but I'll share with you here that their optimal temperature for reproduction is between 77–86°F (25-30°C). At 64°F (18°C) their growth rates is decreased by 50 percent. At 46–50°F (8–10°C) it decreases by 75 percent, and it stops altogether at 39°F (4°C). They will die off at or below 32°F (0°C) and at or above 120°F (49°C). Because of the reduced reproductive rates in suboptimal tempera- tures, outdoor, springtime cycling is a very slow affair in most parts of North America. And then, once you build your thriving bacteria colony, you will likely not be able to keep them alive through the off-season (based on the temperatures noted above), meaning you will have to start from the begin- ning when the next growing season returns.

The need to keep the bacteria that form the biofilter constantly alive is really the only downside to aquaponic gardening that I see. Nothing is per- fect, right? By comparison, in hydroponic gardening, you can start up your system instantly, whenever it suits you. There is no need to create a biofilter or build up nutrients. You simply scoop nutrients out of a jar and away you go. Most of the microorganisms that are important to soil gardening will go dormant and survive a cold winter, to be reawakened in the spring thaw. Not so with nitrifying bacteria. They die off at freezing temperatures and your biofilter will need to be recreated in the spring from scratch.

Grow indoors

I know many aquaponic gardeners who have set up systems in their homes. They use basements, heated garages and rooms formerly occupied by kids who have left the nest. One I recently heard about has the grow beds in the main floor laundry room, the fish tank in the basement and the plumbing to connect them running through the laundry chute!

Your personal circumstances and the climate in which you live may dictate that your only choice to have and enjoy an aquaponics system is to put one inside your house. You will not be alone. Many very successful aquaponic gardeners grow exclusively indoors. However, if you are going to grow indoors, please consider the following before taking the plunge:

Weight — Water weighs about 8 pounds per gallon (1 kg per liter). Gravel weighs about 105 pounds per cubic foot (1682 kg/m³). If you construct a simple aquaponic system using a 100-gallon (380-liter) stock tank as a fish tank (38 lb/17 kg, without water) and two 50-gallon (190-liter) stock tanks as grow beds (25 lb/11 kg each), you will have 838 pounds (380 kg) of weight with the fish tank and 1,457 lb (660 kg) of weight with the grow beds before plants and water during a flood cycle. Between the two of them that's more than one ton. While most gardeners won't place it all in the same space, weight is an important consideration. Be cognizant of the floor structure supporting your tank and your grow beds. Make sure it can handle this kind of weight so you don't end up with your aquaponics system on a lower floor than the one you started on!

Humidity — Aquaponics is a water-based system, typically with warmed water constantly moving and aerating the fish tanks. This will undoubtedly introduce extra humidity into the surrounding environment. This could be considered a benefit if you live in a dry environment or have furnace heat that tends to dry the air in your home over the winter. But be aware of the potential for generating serious humidity if you have a large, open fish tank and plan to heat the water to a temperature that is greater than the

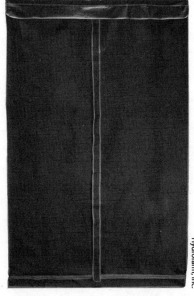

Left: *Mylar grow room open.*

Right: *Mylar grow room closed.*

ambient air. Be especially aware if you have some single-pane windows and cold weather outside. The warm, humid air hitting the cold glass could create a rain forest.

Water spillage — Where there are large quantities of water, there will be spillage. You can count on it. Whether it comes from the water you use to top off your tanks or a freak accident as a hapless aquaponic gardener experienced below. Just be aware that something could happen. Put safeguards in place to prevent it and be prepared in case it does.

Light — Unless you have a glassed-in sunroom with skylights, you will need supplemental lighting of some kind. Today's plant grow-lighting incorporates very sophisticated technology. We provide lots of information about this below, but there are two environmental ramifications to keep in mind now as you decide whether or not to grow indoors and choose your lighting system. The ramifications are heat and power draw. The best, most flexible lighting setup you can buy is an HID (high-intensity discharge) system. The tradeoff is that it generates significant heat and draws significant power. Again, be aware and make your decision with your eyes wide open.

Sound — Finally there is the sound of the water. Your system will pulse with the lively sounds of water moving back and forth between the fish tank and the grow bed. For some, this is a delightful, soothing benefit; to others, it is annoying. Just be aware of how you feel about water sounds before you set up an aquaponics system inside your home in a spot where you can hear it.

One way to mitigate some of these issues is to establish a grow room within another room. This is often done with cube-shaped, temporary structures with zippered doorways and reflective material, such as Mylar, lining the interior walls. I've seen these set up very effectively in basements and garages and they can really help with insect control, lighting effectiveness, sound dampening and water containment.

Grow outdoors in the summer and bring our aquaponics system indoors for the winter (or summer)

You could combine option one (grow outdoors) with option two (grow indoors) and bring your aquaponics system indoors when the weather becomes incompatible with growing outdoors. The obvious issue here is the practicality of moving all that weight and mass around (remember the one

Aquaponics update #18 — Disaster

Last night I fell asleep on my couch. I woke up early to feed my fish and to cook (I have a huge job today). When I went in my room, my tank was almost completely emptied of water!!

I'm still not sure what happened and I am only taking the time to write this because I have to wait for the pet store to open at 9 so I can buy dechlorinator and add emergency water to my tank.

The tank didn't break; it looks like one of the plant leaves may have diverted the water from the filter out of the tank — but I'm really not sure. I don't know if my fish will survive; even my clam was so upset he was out and moving around, which clams don't do.

I have to fill it part way with water once I get dechlorinator — I put the ten-gallon tank in the 55-gallon tank just in case it's the tank that is leaking — threw in the air stone, and hope the fish are fine until late tonight, when I get back from my job, with only air and not a filter.

I am so bummed right now. On the upside, even though my carpet was wet, it was just one small area, and most of the water went into my dresser (tank was on top) and was soaked up by my clothes — so no damage, I just have to do a lot of laundry ... but still, uurrrgghhh.

— Chef Ricky, Bourgeouisie Brunches
December 18, 2010

ton calculation above?). If, however, you design your system to be transported from the onset it is absolutely doable.

Portability was a key design criterion when we created our AquaBundance aquaponics system. We wanted to design a compact setup that could be left outdoors, perhaps on a deck, for the summer then moved indoors for the winter. To accomplish this we made sure that the entire system was no more than 28 inches (70 cm) wide so it could fit through any standard exterior doorway and mounted industrial castors on the grow bed stand so it could be rolled indoors. While the water in the fish tank should be relocated prior to moving the tank, both the grow bed and the fish tank have substantial handgrips on either end for lifting ease. Consider these points if you decide to create your own indoor/outdoor aquaponics system, or purchase a ready-made, turnkey system.

Erect a greenhouse

If you can afford it, and can get whatever permission you might need from your municipality to build it, a greenhouse is an ideal home for your aquaponics system. In fact, in my fantasy dream house, I would have an attached greenhouse off my kitchen. Here's why.

- Build the environment around the plants and fish you want to grow — Greenhouses give you total control over your growing environment. Even seasonality can be removed by adjusting the day and night temperature and adding lights. Want to grow tomatoes in the winter? No problem. — Prefer a lower-energy growing solution? Grow seasonal, cold-weather crops and fish during the winter and forget the lights.
- Maximize, or diminish, sunlight — Greenhouses are designed to maximize sun exposure, but can also be outfitted with a shade cloth to diminish the sun's intensity during the summer months.
- No worries about water — In a greenhouse you can have a hose, hopefully with a dechlorinating filter, for filling up tanks, blasting off bugs, cleaning out clogged pipes and many other tasks. You can use water as you need to without concern for the effect on the floors or walls.
- A special oasis — As I was writing this I got a phone call from a friend who had been in a car accident. He was broadsided by another car, and while he suffered no injuries his car was not so lucky. I asked where he was

now and he said he was in his aquaponics greenhouse. He went there to calm down and relax. Ask anyone who has an active greenhouse, especially an aquaponics greenhouse, and they will tell you what a special place of renewal it is for them. It fills every sense with life. The smell of basil and mint, the feel of the moisture in the air, the sound of water, the taste of a fresh-picked strawberry, the vision of healthy plants and the scrimmage of hungry fish. It is a magical place.

Sylvia in greenhouse.

Greenhouses can be as elaborate and expensive, or as do-it-yourself, as befits both your wallet and your personality. While it is beyond the scope of this book to go into much detail about greenhouse construction (I've listed several good resources in the back of the book), I have learned a thing or two about building sustainable greenhouses that I'd like to pass along in case you decide to embark on a building project.

- Before anchoring your greenhouse with a foundation or footers be sure to surround the exterior with insulation down to the frost line for your climate. This will prevent warm air from seeping out through your floor during the winter.
- Very little sunlight will come through your north-facing wall (or south-facing wall, if you live in the southern hemisphere). You are better off insulating it thoroughly than hoping to get more sun exposure through it. You can also attach it to the side of your home, if that works for you.
- Your main sunlight will enter through your south-facing wall (or north-facing, in the southern hemisphere). That is where you should have either glass or other glazing material, such as polycarbonate, fiberglass or greenhouse film. Whatever you choose, be sure to use a double layer of it with an air space in between to slow down heat loss.
- Consider using salvaged materials, such as discarded double-pane windows and glass sliding doors, and building your greenhouse around whatever is available to you secondhand.

And here are a few observations that will make your greenhouse more aquaponics friendly.

- Built in power and water — Aquaponics systems require a ready supply of both water and power. You won't be happy for long if you are carrying buckets of water and running extension cords out to your greenhouse.
- Reliable power — The power must be reliable and constant; otherwise your fish will quickly die from lack of oxygen. Solar and wind energy are terrific sources of power, but they must be backed up by a battery, a generator or AC power. Even if your grid power is generally reliable, every power source goes down sometimes. We have extremely dependable electrical power here, but last summer there was a fire nearby and for the safety of the firefighters they shut off the power to most of our city for most of a Friday. It happens, so you need to have a plan in place for when it does. Do you have a generator? Can you wire an aerator to a car battery? You may want to look for backup aerators used for keeping minnows alive; you might find some at a sport fishing store.
- Notification of power failure — How will you know if the power has gone down in your greenhouse, especially in the middle of the night? We wish we had a phone line in our greenhouse so we could attach an inexpensive device designed to make a phone call when the power goes down. There are also wireless internet devices that do the same thing.
- Consider water heating — The most power-hungry part of running an aquaponic greenhouse can be heating the fish tank water. Consider that 1 watt is equal to 3.4 BTUs, with a BTU (British thermal unit) being defined as the energy needed to raise the temperature of one pound (.45 kg) of water by 1 °F (0.56 °C). Since a gallon of water weighs about 8.3 pounds, it takes about 2.4 watts to heat one gallon of water 1 °F. So you will need roughly 240 watts to raise 100 gallons 1 °F (or 240 watts to raise 380 liters of water 0.56 °C). If you happened to have a water heater rated at 400 watts, the rating actually means that it produces 400 watts per hour. Thus, if you were to use your 400-watt heater, you would raise your 100 gallons (380 liters) of water 1 °F in about 36 minutes.

All sustainable greenhouse designs now use the notion of a heat sink. According to Wikipedia, "A heat sink is an object that transfers thermal energy from a higher temperature to a lower temperature fluid medium.

The fluid medium is frequently air, but can also be water." You can create a heat sink in a greenhouse by storing a large amount of water on the north-facing, hopefully insulated, wall. Radiant heat that accumulates during the day helps heat the water, like charging a battery. Then at night, the heat is released back into the air as the greenhouse cools down. Whenever this is described to me, I instantly think "fish tank!" The perfect heat sink is a large tank full of a warm water fish in your greenhouse. But be sure to plan to have it on the north-facing wall so that the warm air that is released around the fish tank will be pulled across the greenhouse toward the cool air coming in from the south-facing, non-insulated wall.

• Consider systems design — Finally, you should think through the design of your aquaponics system and specifically whether you want to follow a plan that puts the top of your fish tank higher than your grow beds (e.g., CHOP) or lower (e.g., Basic Flood and Drain or Two Pump Design). If you opt for the fish tank or sump tank to be the lowest part of your system design, you might want to sink it into the ground during greenhouse

My automated aquaponic greenhouse

My aquaponic greenhouse utilizes technology to operate. It's not just a couple of temperature sensors, but a full-blown automation and monitoring system which is the entire nerve center to my highly productive system. It ensures the plants and fish are healthy and growing at their best.

I will admit that it's not necessary to have a hi-tech system for aquaponics, but it does remove some of the tedious tasks involved with the daily operations. Being inside a greenhouse, it doesn't take very long once the sun comes up in the morning for the temperature to jump high enough to kill the plants. A simple temperature monitor automatically opens and closes the vents throughout the day, helping to maintain a more constant temperature. Another module of the controller compares the water temperature to the ceiling temperature during the colder months. If the water is too cold, a circulator pump turns on, extracting some of the heat near the ceiling and storing it in the water. This helps to keep the fish warmer and increases the thermal mass to reduce the amount of heat needed during the night.

Space is very limited inside the greenhouse. I have six grow beds of various sizes. If they were all to fill at the same time, I would simply run out of water. The controller system acts as a timer and runs a sequencing valve that allows the beds to fill one at a time. Since each bed varies in size, the timer is programmed with the appropriate ☞

construction. This will both help insulate the water and also help you to place the grow beds at a more comfortable height. Just be careful not to place the fish tank so low that it is dangerous for small children or an open invitation for your dog to go fishing!

Another nice benefit of having a greenhouse run by aquaponics is that the fish generate what the plants "breathe," i.e., CO_2. That is good for the plants. In intense, environmentally controlled greenhouses, CO_2 is either piped in or generated to supplement the air and boost plant growth. In an aquaponic greenhouse, that probably isn't necessary. The fish do that work for you.

Even in temperate climates…

As Murray indicated earlier, our friends in paradise must also deal with some grumpiness from Mother Nature. For example, torrential rainfall can flood your systems and thereby put them out of balance, by enabling your fish to jump over the side of your now over-full tanks or causing your pH to go

filling times minimizing wasted electricity from overfilling the beds. Another energy savings is that the grow bed pump will shut down during the night — of course, there is a light sensor mounted in the greenhouse to monitor the light levels!

One of the more entertaining modules is the automatic fish feeder. At set times during the day, the fish get fed automatically. The unique part about this feeder is that the amount of food they receive increases or decreases depending on the water temperature. This helps to ensure that they don't get overfed since their metabolism changes with the temperature. It also allows me to be away without worrying about hungry fish. However, I still like running out to the greenhouse at the set times to watch the pellets drop into the water and the fish swarm around the food.

The entire controller system is a series of interface cards that are capable of reading temperature, light, water flow and water level. There is even a camera. There are relay cards that are used to turn on the various pumps, open and close vents and run the heaters during the winter. These interface cards are connected to a central computer that collects and transmits all the data over the internet to a central web server. From the comfort of my desk at work or couch at home, I can monitor all the vital statistics and make adjustments to my aquaponic system.

— Rob Torcellini, Bigelow Brook Farm, LLC

out of range. You may be able to prevent these problems by erecting a cover over your aquaponics setup with corrugated roofing material, to channel the water into a rainwater catchment basin for use in drier times.

Insect control can also be much more of an issue in a temperate climate (although if you have ever had a greenhouse full of white fly you might want to question this). I'm told you can mitigate this by hanging insect netting material from the roof you built above.

Lights

If you are growing in your home, chances are you need to either supplement your window lighting, or grow entirely with grow lights. Even in a green-house, you might choose to supplement the shortened natural daylight of the winter months with grow lights that come on for part of the evening. This allows you to grow not only the herbs and greens that do well with shorter days, but also grow the fruiting plants that crave the longer days of summer.

When you select and run lights for your plants consider the following:

- Duration — As I said above, how long you run your lights will depend on the day-length requirements of your plants. Are they fruiting (e.g., tomatoes) or foliage (e.g., lettuce) plants?
- Spectrum — Be sure that you are covering the blue and red spectrums that are absorbed by plants. You will know this by the language used to describe the lights. If it says "grow light," "plant light," "full spectrum light" or "daylight spectrum" it will work for indoor plant growing.
- Canopy penetration — The light should penetrate deep enough into the plant's "canopy" (structure from top to bottom) to effectively get light to the lower leaves. A mature tomato plant has a very different canopy depth than a lettuce plant.
- Heat — Some light sources produce a lot of heat. This can be a good thing when you are trying to heat a greenhouse in the winter, but not such a good thing if you are growing in a small, enclosed room within your house.
- Energy input for light output — Some fixtures require a lot of power, others are fairly energy efficient.
- Cost — Consider all the costs, including the original cost of the fixture, sized to grow the plants you want to grow, the cost of the replacement

bulbs, the frequency of replacement and the cost of the power to run the light.

Below, I evaluate the three main types of lighting options favored by indoor gardeners using the parameters listed above.

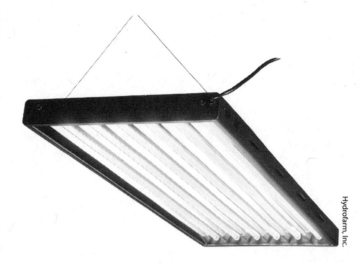

Fluorescent lighting using T5 bulbs

T5s are used in indoor lighting fixtures because they are skinny, so you can pack a lot of them into a fixture. As the number next to the "T" goes down, so does the diameter of the bulb (T12s are bigger than T8s which are bigger than T5s).

Pros of T5s — They have a broad plant lighting spectrum so they work for both fruiting and foliage plants. They are also low power consumers and don't throw off a lot of heat. Plus there are now some "designer" fixtures for them that make them more attractive.

Cons of T5s — They will only reach through 18 inches (45 cm) of plant canopy. For taller plants (again, think tomatoes) you can hang them sideways. Also, their performance drops off significantly after six months, even though they still look just as bright as the day you got them. Thus, you should replace the bulbs after every six months of use.

T5 fluorescent lights.

HID (high-intensity discharge) lighting

These are serious lights for serious indoor growers. They come in five parts:

- The bulb is either metal halide (MH), which is a mostly blue spectrum bulb for vegetative growth, or high-pressure sodium (HPS), which is a mostly red spectrum bulb for the mature, productive stage of a fruiting plant's life.

- The ballast provides the power to the light. It comes in a switchable format that can run either an HPS or an MH bulb.
- The cord set is the power cord that plugs into the ballast at one end; the other end provides the socket for the bulb.
- The reflector is a covering that goes over the top of the bulb and directs the light down onto the plants.
- The fan and vent are often part of HID lights, but these are generally optional. They direct the heat away from the light and plants if that is necessary.

Hydrofarm, Inc.

HID light ballast.

You can purchase each of these components separately if you want a very customized or high-end light. They also come packaged together as a convenient turnkey kit.

Pros of HID — They provide much more intense light that goes through almost any plant canopy. If you get a switchable ballast, you can easily move from a metal halide bulb (blue spectrum for vegetative growth) to a high-pressure sodium bulb (red spectrum for fruiting) for an even more precise plant spectrum. Bulbs last at least a year.

Cons of HID — The bulbs are expensive, they draw more power than fluorescents and they throw off some serious heat.

You might find the chart on the next page helpful to figure out coverage if you decide to go down the HID lighting route.

LED lighting

This is one of the newest lighting technologies on the market. LED lights use solid-state light-emitting diodes, hence their name.

Pros of LED — No heat and very low power consumption, so you never replace the bulbs.

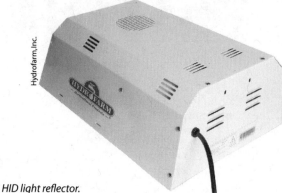

Hydrofarm, Inc.

HID light reflector.

Cons of LED — Because these are still new, I have not seen enough data to be confident that they will grow as well as either of the other two options, and they are still relatively expensive at this time. Also they typically cast a very odd light over your plants and give a red glow to the room. A friend who has just started growing with LEDs uses a flashlight to see what her plants truly look like.

Below is another interesting chart I ran across recently that shows the light output of some different options:

System location and environment — conclusion

Unless you are fortunate enough to live in a year-round growing climate, you will need to design elements of your aquaponic system to account for your local environment. Can you afford a greenhouse? Can you convince your spouse or roommate to donate a room

Top: *Chart of HID light reach.*
Bottom: *Light type comparison chart.*

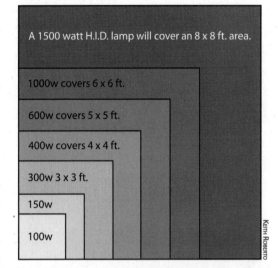

A 1500 watt H.I.D. lamp will cover an 8 x 8 ft. area.

1000w covers 6 x 6 ft.

600w covers 5 x 5 ft.

400w covers 4 x 4 ft.

300w 3 x 3 ft.

150w

100w

Keith Roberto

LIGHT OUTPUT COMPARISON CHART

The lumen output is the same for each of these five scenarios shown below.
Each represents a total lumen output of approximately 50,000.

High Intensity Discharge	T5 Fluorescent	Self Ballasted High Wattage Fluorescent	T12 Fluorescent	Incandescent
1	**10**	**5.5**	**42**	**84**
400 watt HPS H.I.D.	4 ft - 54 watt T5 H.O. Fluorescent	125 watt Fluorescent Lamp	4 ft - 40 watt T12 Fluorescent	60 watt Incandescent (Standard household bulbs)

Sunlight Supply® Inc.

or part of the basement to your system? Do you need supplemental lighting? If so, are T5 fluorescents or HIDs more appropriate for your growing situation? The good news is that there is almost certainly some configuration that will work for you.

6

System design

"Fresh clean veggies and sane protein from your own backyard in your own mini-ecosystem: it doesn't get better than that…and you can quote me on that!"

— Ted J. Hill, Riverside, California

Now that you have decided where to put your aquaponics system, let's explore some of the ways that you can design it. I will lead you through why you might choose one configuration over the other. As the design of your system starts coalescing in your mind, I encourage you to get out pencil and paper, or use Google Sketchup or whatever drawing tool you are most comfortable with, to begin to draw out your designs. They may change some as each chapter builds on the last, but this will give you a place for your system to take shape.

This chapter is intended to be a high-level treatment of the most frequently used system configurations in aquaponics. I will go into much more detail about grow bed, fish tank and plumbing specifications in the next two chapters. But first I want you to think about the bigger picture.

All of these configurations will be about media-based growing, but we will conclude by talking about how to add a raft or DWC (deep-water culture) bed onto your media-based system.

Basic flood and drain

Flood and drain systems are the simplest of all configurations to understand and assemble — thus the word "Basic" in the title. This design works well for any beginning system using a 1:1 grow bed volume to fish tank volume, including small aquarium systems.

In a Basic Flood and Drain design, as pictured below, the grow beds are above the fish tank. Water from the fish tank is pumped up into the grow beds, either through the bottom of the bed if you plan to drain through the pump, or over the top of the bed (A). The water is then returned to the fish tank by gravity when the siphon trips or the power to the pump is turned off by a timer (B).

If you are concerned that adding new grow beds will draw the water in your fish tank down to levels that will stress or even endanger your fish, you can always add an indexing (sequencing) valve. With this valve attached to a timed pump, every time your pump goes on, your system will flood a different grow bed. See the Plumbing chapter for more information about using indexing valves.

The benefits of this configuration are that it is straightforward to assemble and maintain and does not require a sump tank. The downside is that as you add grow beds and go from a 1:1 to a 2:1 bed to tank ratio, you will be draining too much water from your fish tank with every flood cycle unless you use an indexing valve. Low water levels may be stressful to your fish.

Adding a sump tank (CHIFT PIST or CHOP)

The next level of sophistication requires the addition of a sump tank (A) as pictured on the next page. Systems with sump tanks return the water from the grow beds via gravity into a sump tank, where the pump resides.

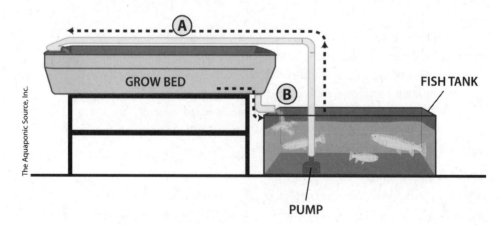

Basic Flood and Drain.

The Aquaponic Source, Inc.

The water from the sump is either continuously pumped up into the fish tank or the pump is activated via a float switch when the water in the sump tank reaches a certain level (B). The water level in the fish tank remains at a constant height because it is always draining out through an overflow pipe (C) back into the grow beds. The water level in the grow beds is controlled via an autosiphon (D). With this approach, water in the fish tank remains at a constant level (the height at which water flows into the overflow pipe in the fish tank) and the fluctuations in water height are relegated to the sump tank. This design is called CHIFT PIST (constant height in fish tank — pump in sump tank) or, as coined by Murray Hallam, CHOP (constant height one pump).

The advantage of these systems is that the fish tank stays at a constant height, which is ideal for the fish. Plus, there is only one pump involved so there is less energy consumed and fewer possible failure points than some of the upcoming designs.

This system works very well but one disadvantage is that it requires that the grow beds be perfectly level to function properly. Hallam recently created a new design, CHOP2, that eliminates this issue. With CHOP2 the pump sends the water from the sump tank to both the grow beds and the main fish tank. The water from the fish tank and grow beds runs back to the central sump. Hallam describes it as "kind of like a double loop water flow with the sump as the central mixing point."

Another disadvantage of a CHOP-based system is that you need to acquire a sump tank that is low enough to reside under your grow beds, but has enough volume to hold more than the volume of water in those beds;

CHOP (constant height, one pump).

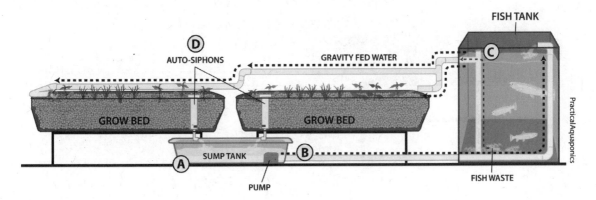

otherwise, when the beds are full, the sump will be empty and no water will be pumped from the sump to the fish tank. Because of the sump tank, you may need more square footage to set up your system. Another possible disadvantage is that the fish tank now needs to be higher than your grow beds so that the fish tank water can drain directly into the grow beds via gravity. There is also a risk that you will burn out your pump and your system will stop functioning if your sump tank water level gets too low.

Adding a second pump

The way to get around the need for the fish tank to be higher than the grow beds is to add a pump to the fish tank.

Configured as shown below, you no longer need to rely on gravity to move the water from your fish tank into your grow beds. Now the pump in your fish tank can either be constantly pumping water into the grow beds, or turned on by a float valve (A). As the autosiphons in the grow beds trip, their water drains into the sump tank (B). The pump in the sump tank is regulated through a float valve (C). When the water in the sump tank reaches a pre-set height, the float valve is triggered, switching on the pump that pumps water back to the fish tank. You can set the trigger heights for the float valve switch (C) such that enough water is left in the sump tank to support a school of fingerlings or another variety of fish.

The advantage of this design is that, while the sump tank still needs to be lower than the grow beds, the fish tank can be whatever height is most convenient for you. The disadvantages are you are now necessarily introducing

2 - pump system.

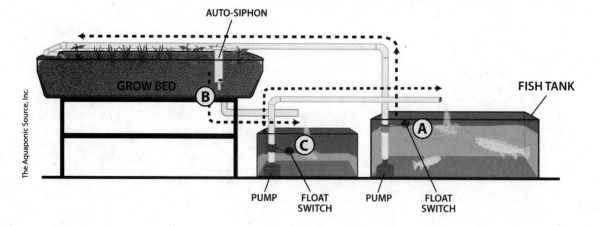

another float valve, which can be fussy, as well as powering a second pump. Also, the water level in your fish tank will fluctuate, though you can control it by how you set the float valve. Finally, this system necessarily switches both pumps on and off, which can shorten the life of the pump.

Barrel-ponics®

Barrel-ponics® is an aquaponics system style invented by American Travis Hughey in 2003. Hughey set out to create a system that could be built out of inexpensive and/or recycled materials found anywhere in the world, with a special focus on Africa. In 2005 he wrote the Barrel-ponics® manual and in 2007 he made it available online as a free download from his website.

This system consists of three primary components, all made from recycled plastic barrels: the flood tank (A), the grow beds (D) and a fish tank (C). The flood tank is a mechanical device (in some ways not unlike the mechanism used in a toilet tank) that operates as a water timer as follows:

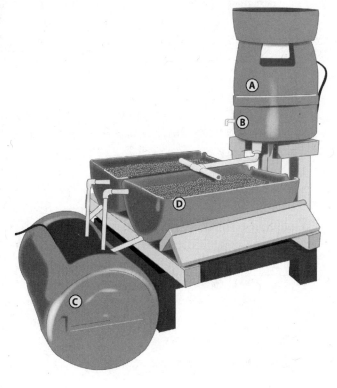

Barrel-ponics® system.

Water from a pump in the fish tank enters the flood tank. When the proper level is achieved, a small siphon begins filling a counterweight which, when heavy enough, pulls the valve (B) open to allow the nutrient rich water from the fish tank to flood the gravel- (or other media) filled grow beds (made from barrels, of course). When the water has drained from the flood tank, the small siphon stops and the counterweight begins draining (into a grow bed) via a small drain hole in the bottom. When the counterweight empties, it becomes light enough for the valve to close and the cycle begins again. Nutrient rich water in the grow beds slowly drains back into the fish tank through a restricted outlet. The entire system is designed to return all water to the fish tank in case of any malfunction via an overflow tube in the flood tank and piping in the grow beds.

Some of the advantages of this system are that can it be made from recycled and/or very inexpensive materials that are widely available; the flood characteristics are easily changed as needed; the pump runs continuously, improving aeration and extending pump life; and the flood tank can operate on flow rates as low as 10 gallons per hour (gph), allowing it to function on low power and in minimal situations. This design has been detailed in Hughey's instructional guide entitled Barrel-ponics® (Hughey, 2011); it is

Raychel's aquaponics story

There is a line from a song that says "You make my heart sing." I can truly say that aquaponics makes my HEART sing.

I am 70 years old and have always been a farmer. I grew up on a 280-acre farm in Iowa. I was sure that was not where I wanted to live. I was sent to Hawaii when I was in the navy and I never left. I bought an acre of land and went to work gardening and farming. I love growing food. My heart leaps when I see those shoots coming out of the ground. You go out each day to see how much has grown. You get totally excited when you see the first fruit and cannot begin to explain the joy when you take the first bite.

I planted a fruit farm in 1999 and sold fruit at the swap meet for 10 years but the weekly journey became too much. Then one day my chiropractor came running over to the lab where I work with an aquaponics video. I was hooked from that moment on. Because I am a scientist and a researcher I began to search the internet. I set up a little system with a fish tank and six half barrels. Wow … it worked! I was so excited. Things went well for a while, but then I developed a torn retina

and had to quit. I was only allowed to watch TV. No reading and no computer for six months. I about died from boredom. Then my chiropractor came running over to tell me about aquaponics classes in Honokaha HI. I had to wait for the next class but I was there. I left that class so filled with expectation, I was about to burst. I went home and built two very large, on-the-ground rafts and began to grow food again.

Aquaponics has given me a reason to sing. It has made my life productive. I can do what it takes to keep it going. It gives me joy to feed the fish each day and I even give the tanks names. Each tank is different. I yell over the fact that the cucumbers are growing and producing. Aquaponics gives me a chance to share with people not only the vegetables but the knowledge of how to make their own system and become sustainable. Someone else said that it is hard to put the feeling into words. The Hawaiians have a saying: "It gives me chicken skin." I guess that about sums it up. Aquaponics makes my heart sing and gives me chicken skin.

— Raychel A Watkins, Waianae, Hawaii

widely followed throughout the world and supported by a large group on Yahoo. You will have plenty of company if you choose this approach.

One disadvantage is that this system is based on using 55-gallon (210-liter) plastic storage drums, preferably in blue. Some may not find this approach attractive (although the barrels can be concealed with a good-looking cover) and support framing can be somewhat more complex.

In my opinion these systems are good for first-time aquaponic gardeners who are looking to start small and inexpensive and wish to expand later, or for use in third world environments, as Hughey originally envisioned them. The barrel halves also make excellent grow beds for larger, more complex installations.

Hybrid system

I stressed at the beginning of this book that I am only going to discuss media-based, backyard systems. The main reasons for this are that raft (DWC) and NFT (nutrient film technique) style systems are more limited in what they can grow and both require solids filtration.

Well, now I'm going to break my own rule and talk about why you might want to consider creating a hybrid aquaponics system that includes both media beds and either raft or NFT systems.

Let's face it. While I stand by my assertion that media-based systems grow the greatest variety of plants, including indeterminate tomato vines and even banana trees, and grow them well, harvesting can be a bit of a pain when it comes to pulling plants entirely from the media. In order to reuse the media and compost the plant roots, you must separate the media from the plant roots. This isn't a big deal when you are talking about a pepper plant that will produce for a year, but salad greens can be harvestable in two months. And if your family is like mine you might enjoy salads four or five times a week. That is a lot of tedious root cleaning.

In a raft or NFT system, foliage plants can be started in grow cubes, then transplanted straight into the raft or channel at 28 days. Because their roots hang freely in the water, you harvest them by simply pulling them through the growing hole, cutting off and composting the root mass and then enjoying the fresh produce.

But what about filtration? I have great news. Your media bed is a solids filter, right? If you first run the fish tank water through your media bed, then drain it out to a raft bed or NFT channel, you will have done most (read on to understand the implications of "most") of the necessary filtration without clarifiers, degassers or mineralization tanks. Best of all, you keep those wonderful fish solids in your ecosystem, feeding your worms and providing even better nutrition for your plants.

As of this writing, these types of systems are still being developed and tinkered with but early results are excellent. Most are saying, however, that some small amount of solids filtration is still required after the water leaves the media beds. Since you will probably need a sump tank and another pump to move the water from where it leaves the media beds to the entry point of your NFT or raft bed, you can simply collect the fine solids there. One great idea for doing this is to surround the pump with filter material and drop it into a plastic milk crate in the sump tank. Now, as water is drawn into the pump to feed the NFT or raft beds, it passes over the filter pads and deposits the last bits of solid waste. Rinse the pads occasionally to clean them and direct the waste water right back into the main fish tank or media bed for organic breakdown.

System design — conclusion

At this point you should have a rough idea of how you want to design your system. Are you building a small aquarium system? If so, a Basic Flood and Drain style is the most appropriate. Are you designing a large system, but want to keep your power as low as possible? A CHOP system may be for you. Do you have aspirations to add a raft bed at some point? This summary chart might help to clarify your thinking.

	Basic Flood and Drain (FD)	CHOP	2-Pump	Barrel-ponics®	Hybrid
Ease of assembly	H	M	M-L	M	L
Ease of maintenance	H	M	M	M	L
Ease of harvest	M	M	M	M	H
Expandability	M	H	H	L	Depends on tank plumbing configuration
Complexity / Risk of problems	L	M (float valve)	H (2 float valves, 2 pumps)	M (float valve)	H (2 grow styles)
Power use	L	L	M	L	M

In the next chapter we will delve into the details of how to size and acquire the grow bed(s) and fish tank(s) for your new aquaponics system.

Aquaponic System Design Rules of Thumb

I recommend a media bed for new, hobby growers. Why not NFT or deep-water culture (aka raft or DWC)?

- A media bed performs three filtering functions:
  - mechanical (solids removal)
  - mineralization (solids breakdown and return to the water)
  - biofiltration
- Because the media bed also acts as the place for plant growth, it basically does everything all in one component — making it all simple.
- Media also provides better plant support and is more closely related to traditional soil gardening because there is a medium to plant into.
- The cost of building the system is lower because there are fewer components.
- It is easier to understand and learn.
- Basic Flood and Drain is the simplest system to design and is appropriate for a 1:1 grow bed to fish tank volume.
- CHOP or 2-Pump systems have sump tanks and enable a 2:1 and even up to a 3:1 grow bed to fish tank ratio. More grow beds filtering the water is generally better for the health of the fish.

Section 3

The hardware

7

Grow beds and fish tanks

*"To me, aquaponics represents the possibility of growing
life-sustaining nutrient rich foods in a way that can actually positively
affect the health of our planet. Most agricultural methods deplete our
soils, but we hold the magic key with aquaponics. By composting the
waste from our plants and fish, we can nourish the land that we need to
do so much work to repair while leaving it fallow or revitalizing it with
native plants that combat erosion and provide much needed wildlife
food and habitat. My hope is that aquaponics will develop in a
way that allows us to grow all of the food a growing global
population will need while preserving our natural biomes
in all of their beautiful diverse glory."*

— Molly Stanek, Milwaukee, Wisconsin

When writing this book I struggled with whether I should separate
grow beds and fish tanks into their own chapters. In a sense they are
like twin siblings. They are reflections of each other, often made of the same
material, living their lives in the same space bound together through a common purpose. Yet they each have their own identity, set of requirements and
function within the larger whole.

In this chapter we'll talk about the volume relationship between the grow
bed and fish tank and the common requirements of both, and then look at
the unique requirements of each. Next we'll go into the most commonly

used products and materials for grow beds and fish tanks. We'll end with a discussion of vertical gardening systems.

Volume relationship between grow beds and fish tanks

One of the main secrets to creating a successful aquaponics system is to create a balance between the amount of fish waste and the ability of the biofilter and the plants to convert that waste into plant food. Too much waste and the biofilter gets overwhelmed, becomes anaerobic and the fish suffer. Too little waste and there is not enough nutrient to make the plants grow. The biofilter is the colony of bacteria that live on every moist surface of your aquaponics system, but primarily in your grow medium.

What follows is a set of ratios that have emerged from a decade of trial and error in media-based systems around the world. These ratios start with the system input: the fish feed. This creates the fish waste that drives the biofiltration need and, consequently, the surface area of your planting beds.

We make assumptions at every point along the way, based in part on aquaculture science and biology but mostly on aquaponics experience. From these assumptions we have arrived at a standardized set of Aquaponic Gardening Rules of Thumb that I will refer to throughout the rest of this book and have summarized in the appendices.

One of the first Rules of Thumb for a beginning aquaponic gardener is: The total volume of all the grow beds that you connect to a single fish tank should be at least equal to the volume of the fish tank, in order to provide adequate filtration for your fish tank. Driven by this simple 1:1 ratio, plans are scattered across the internet for aquaponic systems built of two halves of a closed container, often a scavenged recycled food shipping or storage container. Cutting one of these containers in half creates a grow bed with a matching fish tank. Or you could use two reclaimed bathtubs, or two 100-gallon (375-liter) stock tanks.

However, once the aquaponics addiction fully takes hold and you are ready for larger systems with 300-, 500- or 1,000-gallon fish tanks, you will find that there is flexibility within that simple ratio. You can safely push it

Fish Feed =>	Fish stocking density (fish + water) =>	Fish waste =>	Biofiltration needs (surface area of the media) =>	Grow bed surface area

to a 2:1 grow bed to fish tank ratio and beyond, which provides even more filtration in your system and is even better for the long-term health of your fish.

The reason why I initially recommend a 1:1 ratio is because it enables the simplest design for the beginning aquaponic gardener: the Basic Flood and Drain style system. As I pointed out in the earlier chapter on system design, in order to pump more water into your grow beds than you would get in a 1:1 system, you need to either add a sump tank or use indexing valves; otherwise you will drain your fish tank too far down on each flood cycle. However, if you are comfortable adding a sump tank or using indexing valves at the beginning of the evolution of your system design, feel free to start with a 2:1 grow bed to fish tank volume.

Common grow bed and fish tank requirements

Because the grow bed and fish tank are joined together through recirculating water, they share many of the same materials requirements. Both must hold water without leaking and must not bow, crack or completely fail under the weight and stresses they will experience (structural considerations), and both must do it in a way that is not toxic to their living inhabitants (toxicity considerations).

Structural considerations

- Waterproof — This may seem obvious, but it is imperative that you retain the water in your system. Watertightness becomes especially important where any plumbing fittings enter and exit your grow bed and fish tank. Become friends with rubber gaskets and marine grade silicone.
- Strong — Water weighs 8.34 lb per gallon, or 1 kg per liter. Expanded clay media weighs approximately 1 lb (450 g) per liter, and gravel can weigh more than twice that. Combine this with the weight of the plants, the force of water continuously filling and draining and the expanding pressure of root growth and you have tremendous lateral strain on your containers. Be sure to select the materials for your bed and tank accordingly. Avoid using hydroponic grow beds and nutrient reservoirs as aquaponic grow beds. They are typically made of thin material and were never meant to hold 12 inches (30 cm) of gravel. Avoid plastic storage containers too, unless you shore them up with a support strap or a frame. Again, you will find that

they are too thin and will not stand up to the weight and other forces that strain an aquaponics container.

Toxicity considerations

- Non-toxic, food safe — As before, this may seem obvious but it is important enough to warrant pointing out. Besides being a source for your food, your aquaponic system will be home for many living things: fish, plants, bacteria and worms. They all depend on you providing a safe, non-toxic environment. This becomes especially important if you are re-purposing storage containers to create your aquaponics system. Be very certain that you know what was formerly stored in those containers! Also be cautious when using plastics that have been made with recycled content. While recycling is clearly earth-friendly, it also exposes you to a loss of control over the exact composition of your container and can't be guaranteed safe for your plants, fish, bacteria and worms.
- Inert — Be sure that whatever materials you choose are inert, e.g., they won't leach anything that will alter the chemical composition of your system. Do not use metal containers, not even galvanized metal, for either the grow bed or the fish tank, unless they are lined. Metals can quickly corrode, throwing your system off-balance by lowering your tank's pH. Metal containers may also leach undesirable chemicals into your system. Copper and zinc are particularly dangerous for fish. Uncoated concrete can also alter pH and cause problems with system balancing, although there are a number of food-safe options for providing a barrier between the concrete and your water.

Special considerations for the grow bed

Let's face it. Grow beds take up the most space and will supply you with the lion's share of the food that your system produces. My advice is to start planning your aquaponics system by planning how much space you have for grow beds and how much food you want to grow. Once you have this figured out, you can evaluate the number and size of fish tank(s) you will need to support them.

The amount of grow bed space you have will govern the size of your fish tank. Here's why. The plants need the fish waste to thrive. The bigger the grow bed space and thus the more plants, the more fish waste required.

Simple — you need enough fish to support your plants. As mentioned earlier in this chapter, the recommended beginning grow bed to fish tank ratio is approximately 1:1, i.e., the fish tank volume should be approximately equal to the volume of the grow bed, but a 2:1 ratio is fine as well as long as you account for the greater water draw from your fish tank in your system design. This ratio can also be thought of in gallons per cubic foot, striving for 7.5 gallons of fish tank to every cubic foot of grow bed (or 1000 liters to every cubic meter). For example, a 50-gallon (225-liter) tank would be able to support about 7 cubic feet (0.225 cubic meters) of grow bed.

The heart of this ratio is based on the amount of waste that the fish produce, which is entirely dependent on how much the fish are fed. It assumes that you will use a complete, high-quality commercial fish feed. It also makes assumptions about how much protein is in the feed. Feed for omnivorous fish (tilapia, cod, catfish, goldfish, koi) is typically lower in protein (32 percent) than feed for carnivorous fish (trout, perch, barramundi; 45–50 percent). We are going to ignore this subtlety for now, except to say that the higher the protein content in the feed, the more waste a pound of fish produces. Finally, these recommendations assume a fish stocking ratio of 0.1–0.2 lb of fish/gallon of tank water or 1 lb of mature fish per 5–10 gallons of tank water (or 1 kg of mature fish per 40–80 liters). A mature, plate-sized (12-inch or 30-cm) tilapia weighs about 1.5 pounds (700 g). They typically achieve this weight in 9 to 12 months depending on how ideal their growing conditions have been and how often you feed them.

The Aquaponic Source, Inc.

AquaBundance grow bed.

Aquaponics grow bed zones

Surface or dry zone (Zone 1) — The first 2 inches (5 cm) is the light penetration and dry zone. Evaporation from the bed is minimized by the existence of a dry zone. This dry zone also protects the plant base against collar rot. Additionally, by ensuring that this zone is kept dry, algae is prevented from forming on the surface of the grow bed media and moisture related plant diseases such as powdery mildew are minimized.

Root zone (Zone 2) — Most root growth and plant activity will occur in the next zone of approximately 6–8 inches (15–20 cm). In this zone, during the drain part of the flood and drain cycle, the water drains away completely, allowing for excellent and very efficient delivery of oxygen rich air to the roots, beneficial bacteria, soil microbes and the resident earth/composting worms.

During the flood part of the cycle, the incoming water distributes moisture, nutrients and incoming solid fish waste particles throughout the growing zone. The worm population does most of its very important work in this zone, breaking down and reducing solid matter and thereby releasing nutrients and minerals to the system. "Worm tea," as it is commonly known, will be evenly mixed and distributed during each flood and drain cycle. "Worm tea" and the fish are entirely compatible.

Solid collection and the mineralization zone (Zone 3) — This is the bottom 2 inches (5 cm) of the grow bed. Fish waste solids and worm castings are finally collected here.

Your comfort is the main consideration for grow-bed width. Think about whether you will be tending your beds from both sides or from only one. The greatest distance that most people are comfortable reaching across is about 30 inches (75 cm). If you are able to tend your bed(s) from both sides then you can extend this to about 48 inches (120 cm) wide and still have easy access.

— Murray Hallam, Practical Aquaponics

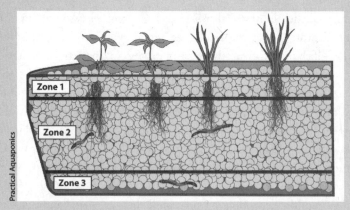

Aquaponics grow bed depth.

Practical Aquaponics

Zone 1

Zone 2

Zone 3

There are 7.5 gallons per cubic foot of volume. Assuming you create 12-inch (30-cm) deep grow beds, (which we will soon cover), you can more simply consider square feet instead of cubic feet. This means that if you have four 4 foot by 8 foot grow beds, you will have 4 x 4 x 8 = 128 sq ft of grow bed space. Now multiply 128 sq ft by 7.5 gallons per cubic ft. If you are using a 1:1 ratio of grow bed to fish tank volume, the result is you will want approximately 960 gallons of fish tank stocked at a mature weight of 1 lb of fish for every 5 gallons of water, or roughly 200 fish if you are growing tilapia. If you design your system using a sump tank or indexing valves you can add twice the number of grow beds (eight vs. four) or cut the size of your fish tank in half for the original number of grow beds (four).

Most experience aquaponics gardeners use grow beds that are at least 12 inches deep (30 cm). While there is no scientific research behind this recommendation, and there are successful aquaponic gardeners using both shallower and deeper grow beds, 12 inches (30 cm) has become the standard. This depth provides enough media density to support most plants and encourages the bacteria in the grow bed to fully establish. A 12-inch (30-cm) deep bed rarely or never needs to be cleaned out because the robust ecosystem enabled by a deep grow bed breaks down the solid waste and takes care of this for you. Below is an excellent explanation by Murray Hallam from *Practical Aquaponics* of grow bed dynamics — reprinted with permission.

The type of plumbing system you choose will affect where and how the water comes in and drains out of your grow bed. However you design your plumbing, you likely will have a pipe coming up through the media. In this case, you will want to block the area around the pipe so that it stays clear of the media. Doing so will make it much easier for you to clean and to prevent roots from growing down the pipes. A 12-inch (30-cm) section of 6-inch PVC works well. Be sure to drill an array of holes all around the pipe on 2-inch (5-cm) centers to allow water to flow through. The holes should be slightly smaller than the smallest pieces

Media guard.

The Aquaponic Source, Inc.

of your media. I've found that a quarter-inch bit works well for Hydroton® and gravel.

Special considerations for the fish tank

Sizing your fish tank defines both the type of fish you can grow and the inherent flexibility of your aquaponics system. Therefore, you should also consider the tank size early in your design process. If you are building a small, desktop system using an aquarium, you will be restricted to aquarium fish that will live comfortably in the size aquarium you own. If you want to grow larger, edible fish like tilapia and trout, there are several additional considerations. The most important rule of thumb is to make sure the tank is made of sturdy, food-grade or food-safe materials. Next, make sure that it is at least 18 inches (46 cm), deep and holds at least 50 gallons (190 liters) of water. Tanks need to hold approximately 50 gallons (190 liters) or more in order to grow "plate-sized" fish (at least 12 inches and 1.5 lb, 30 cm and 680 g). In general, you will see smaller temperature and pH fluctuations with a larger tank than with a smaller tank, and smaller fluctuations are better. A good rule of thumb is to use a 250-gallon (1000-liter) or larger tank if you have sufficient space.

A tank with a round or oval shape is preferable over one with a square or rectangular shape, although both will work in the low stocking densities that hobbyists use. This is because the flow dynamics of round tanks tend to discourage the development of dead zones, i.e., areas in the tank where there is little water flow and exchange or gas exchange or chemical exchange. A round tank with a conical bottom is best of all because the solids will gravitate toward that area and be easily pumped out and into the grow beds. That said, a rectangular or square fish tank is more space efficient and as long as you follow my stocking density recommendation and there is sufficient water movement in the tank agitating and suspending the solids, you shouldn't have a problem.

Another consideration in selecting your fish tank is surface area. If possible, select a fish tank that is short and wide instead of deep and narrow. This is because short, wide tanks have a far higher water surface area to water volume ratio than deep, narrow tanks. The greater the surface area, the better the gas exchange (dissolved oxygen uptake and carbon dioxide removal) for the tank.

Since your fish tank will be extremely heavy and difficult to move once filled, carefully consider where you place it. Be sure it is on a solid surface that can handle the tank's weight when it is filled with water. At 8.3 pounds per gallon, you will want to be careful not to reach a weight that might exceed the structural limits of the surface you are planning to use.

Ideally the fish tank should be located indoors, in a greenhouse or outdoors in the shade. While we often think of needing to heat our fish tanks, water that is too hot can be just as deadly as water that is too cold. Not only do some fish (e.g., trout) require cool water, but as water temperature rises it becomes less and less able to hold oxygen. While there are several factors that influence exactly when water becomes "deaerated," such as altitude and salinity, it is generally safe to say that water that is warmer than 100 °F (38 °C) can no longer hold enough soluble oxygen to support fish and nitrifying bacteria. It is much easier to heat a tank full of water than it is to cool one, so if you have to choose between the two, choose to heat your water when necessary.

Wherever you choose to set up your tank, you will be well served to at least partially cover it to help prevent debris, children and pets from falling in. Covering it will also lower the amount of light reaching the tank. Fish don't require sunlight to thrive and the extra algae growth from sunlight could become a problem. Plus, fish can be skittish and seem to appreciate having a covered area in the tank where they feel they can hide from predators.

Stock tank on car.

Finally, I once read a suggestion that a net or cage be designed right into a larger fish tank to make it easier to raise the fish and catch them. While not exactly sporting, I think this is an excellent idea and one that I wish I had incorporated into my own aquaponic fish tanks!

Commonly used products and materials

As long as you follow the structural and non-toxicity guidelines outlined at the beginning of this chapter, there is

RickOpersteny,SminStudiosAquaponicDesign

almost no limit to what you can use to build your grow bed and fish tank. The following is a list of the most commonly used products and materials.

Stock tanks

Guidelines — Next to the grow bed and fish tanks we manufacture, this is my favorite source. If you do an internet search on "stock tanks" you will quickly discover that there are two kinds: plastic, dominated by Rubbermaid, and galvanized steel. While galvanized steel can be coated or lined to prevent rusting and leaching of toxic metals, I would encourage you to lean toward the plastic versions. They are sturdy, reasonably lightweight and have been designed to hold water in an outdoor environment without rusting or leaching.

I've used a 300-gallon Rubbermaid tank as the centerpiece of my aquaponics system for years, and it has worked out very well for me. I know many people who have used 50–60-gallon (200-liter), 12–14-inch (30–35-cm) tall stock tanks as grow beds with great success.

Because something this big can be very expensive to ship, see if you can find them through a local feed and grain or ranch supplier before turning to the internet.

Pros — Because these have been designed to hold water in an outdoor environment, they are very sturdy and require no extra support. They are manufactured in dimensions that work well for both fish tanks and grow beds, and depending on where you live, they may be easily available.

Cons — You can generally only find these tanks through stores by purchasing them new. There is therefore an expense and some will not want to introduce more plastic products into the world. I've come to terms with this by balancing it with all the ecological benefits aquaponics brings. You will need to reach your own conclusion. These tanks are also large, to very large, containers that can be challenging to get home and expensive to have delivered. Last, aesthetics are not a strength for these tanks so if looks matter to you, you may wish to consider some alternatives.

55-gallon (210-liter) storage barrels or drums

Guidelines — Travis Hughey has popularized the use of plastic storage barrels through his Barrel-ponics® manual (see Resources). These barrels are generally 23 inches (60 cm) in diameter by 35 inches (90 cm) tall and typically come in three colors — white, blue and black. You should insist on knowing what was last stored in them and be sure that it wasn't anything that would be toxic to the fish. Hughey suggests using only blue and white barrels and avoiding black ones because they are most often used for storing chemicals, which can be toxic. He prefers blue barrels over white because the white attracts algae.

Complete small aquaponics systems can be created with one, two or three barrels. To create a grow bed and fish tank from one barrel, cut it around its circumference 12 inches (30 cm) from the top, creating a 12–inch-deep (30-cm) grow bed and a 23–inch-deep (60-cm), 18-gallon (70-liter) fish tank.

Barrel grow bed, barrel systems designed by Coastview Aquaponics.

To use two barrels you should cut a 13 x 23 inch (30 x 60 cm) panel out of the side of one of the barrels, leaving 6 inches (15 cm) between the cut line of the panel and the top and bottom of the barrel. This becomes your fish tank. Next split the second barrel in half lengthwise to create two grow beds.

The Barrel-ponics® design guide uses three barrels. I defer to that guide for those detailed instructions.

You can also use barrels just for your grow beds and one of the other options I have listed for your fish tank. Be sure to create supports for the sides of the grow bed to prevent them from bowing due to the weight of the gravel.

Pros — This is an excellent example of recycling plastic materials that might otherwise end up in a landfill. The barrels are easy to find and inexpensive or even free making them a viable choice for the beginning and space-constrained aquaponic gardener.

Cons — Because these barrels are made of flexible plastic they need to be framed on the sides to ensure they are strong enough to hold the weight of the media over the long run. You can't always be 100 percent sure of what was stored in the barrels so there could be some risk of contamination. Also, they aren't very attractive, so if you are designing a system for your home you might want to consider alternatives. Finally, you will be limited to the

55-gallon (210-liter) size of the barrel for the fish tank, unless you just use the barrels for grow beds and figure out another solution for a larger fish tank.

IBC totes

Guidelines — IBC (intermediate bulk container) totes are plastic shipping containers surrounded by metal frames that are typically used for transporting liquids. They come in a variety of sizes, but the two most prevalent have capacities of 275 gallons (1000 liters) and 330 gallons (1250 liters).

IBC totes are ideal for building aquaponics systems because they are watertight, made of fish-safe materials and come with their own steel frames. Because they are generally not exact cubes but rather rectangular, the grow bed can be made out of the top 14 inches (35 cm) (that is, where the metal frame on the top of the tote will most easily be cut and removed), with the bottom section becoming the fish tank. The grow bed then can sit directly on top of the back half of the fish tank. You can even make this design attractive and more functional by framing it in wood and putting the grow bed on rolling tracks mounted on the fish tank. (King, 2011)

The downside of this design is if you fill the entire fish tank with water and stock it according to the stocking density generally recommended in this book, you will have only half the minimum amount of grow bed (i.e., biofilter) volume you need (1:2 vs. the recommended 1:1). You can get around this by only stocking the fish half as densely, or by using 1 lb of fish per 10 to 15 gallons of water (or 1 kg of fish per 85 to 125 liters of water).

You can also use multiple totes to build larger systems, including using three beds to create a system with one tote as a dedicated 275-gallon (1000-liter) fish tank and the other two carved up into three grow beds, plus a sump tank. Then follow the instructions for building a CHOP system found in the earlier Systems chapter of this book. Murray Hallam shows how to build this system in his *DIY Aquaponics* video, and a kit for building it is also available from The Aquaponic Source.

Used IBC totes are relatively inexpensive (usually from $50–100 US) and can readily be found through eBay®, Craigslist or other online sources, or in the Yellow Pages under "Barrels and Drums — Plastic." Your best bet is to find a local source if you can, because shipping will probably cost more than the tote itself.

Practical Aquaponics

IBC Tote system ("toteponics").

Just as with the storage barrels described earlier, be very careful about what was previously stored in an IBC tote. They are often used to ship a wide range of nasty chemicals that are extremely difficult to completely clean out of your system and that could jeopardize its health. Do not buy a used IBC tote if you don't know what it was used to carry.

Totes come in different styles. Some have pipes on the tank bottom that could obstruct your water flow. If you possibly can, look inside the tank before buying it and make sure it will work with your design. Also, we have found that the top of an IBC tote is generally not watertight because they are designed to let some air out to relieve pressure that could build up in the tank during shipping. Be sure to test your tote grow beds for leaks before adding media, and seal any problem areas with marine-grade silicone.

Unfortunately the plastic used in the totes is not UV stable so it can become brittle and eventually crack if left in outdoor sunlight. Here are the options Murray Hallam recommends: "The best way to paint your IBC totes is with an automotive plastic primer, then use an industrial grade spray enamel as a top coat. This does a really good job, but is expensive. The other way is to use a wipe-on/wipe-off primer, then paint the tote with a water-based exterior house paint. This is much lower cost and avoids spraying, but the paint comes off fairly easily when bumped."

Pros — IBC totes are a very inexpensive way to create both a fish tank and a grow bed out of a single container. They come with their own frames so all you need to do is remove the top and add the plumbing. Finally, they can be made attractive with some framing and paint.

Cons — Like the barrels described above, IBC totes are sometimes used to transport toxic materials. You might not always know the full history of the container you purchase. The plastic that the containers are made of is not UV stable, so if your system is going outside you will need to paint it. Also, cutting through the metal frame can be quite a chore unless you have the right tools. Finally, if you build a system out of one IBC tote the grow bed will only be half the volume needed to properly filter the fish tank so you must keep your fish stocking density very low.

Flexible pond liner and "dead" vinyl

Guidelines — Pond liner is waterproof material that comes in large, flexible sheets and is used to create man-made ponds. You can cut it with a utility knife or even kitchen scissors and bond it with adhesive designed for the job. Using pond liner will give you more design flexibility than any other option listed here. You can convert just about any sturdy container into a fish tank as long as it is at least 18 inches (45 cm) tall, or into a grow bed as long as it is at least 12 inches (30 cm) tall. Just be careful how you cut the holes for your drains — cutting them too big will require patching the liner.

Most liners are either EPDM (ethylene propylene diene monomer) rubber or PVC (polyvinyl chloride). While more expensive, EPDM is the better choice if you can afford it because of its

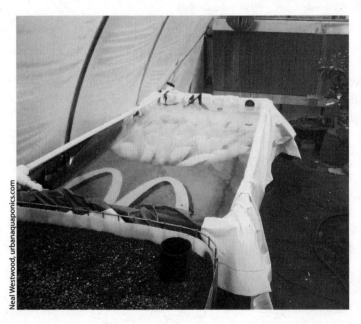

McDonald's, dead vinyl.

Neal Westwood, urbanaquaponics.com

durability and lack of plasticizers. Because EPDM liners are rubber-based, they are extremely flexible (much more so than PVC liners), and they do not contain any plasticizers that can make the liner brittle and crack with age. The lack of plasticizers removes any question about being fish-safe.

Another option for liners is "dead vinyl." This is the name given to recycled billboard material. It is watertight (although you should check for cracks and leaks) and readily available through signage companies. While it won't last nearly as long as a good EPDM pond liner, it can generally be found for free.

Pros — Pond liner gives you absolute flexibility in the design of your beds and/or tanks. It is also compact so you can order it online and have it delivered. Much easier than transporting a 300-gallon tank on the roof of your car!

Cons — The biggest problem with using pond liner or dead vinyl is not with the liner itself, but rather with the frames or containers that you may use to support it. Wood is the most commonly used material for framing and in an aquaponic environment wood can be subject to termites and rotting. Also, it does not give a very finished look unless you somehow seal the edge of the liner to the container. Water can get between the two and cause a problem. With dead vinyl you have the added element of possible toxicity through the ink leaching on the print side, and the glue from the non-print side. As with any other recycled materials, the key to safe use is to ask lots of questions.

Aquariums

Guidelines — Clearly I'm just talking about the fish tank here. Aquariums are great for small or beginner setups, as hospital tanks for isolating ailing fish and for separating the young fry from aggressive older fish. They are

Closet system.

Rob Torcellini, Bigelow Brook Farms, LLC

also ideal for creating a beautiful indoor environment and for those who want to view and interact with their fish rather than just eat them.

Pros — The best option for small indoor setups and for gardeners who want to see their fish. Widely available through pet stores and aquarium supply stores. They are also easy to find used online, and at garage sales and flea markets.

Cons — Strictly recommended for indoor use because a glass aquarium can be subject to wide temperature swings because their clear, thin glass provides very little protection against heat loss at night and enables substantial heat gain in direct sunlight during the day. As an example, I came inside at 7:45 a.m. on a warm September morning after feeding the fish and checking our systems. We had a glass tank which I was using as a nursery tank for our fry fish alongside one of our AquaBundance tanks on our south-facing deck. The glass tank was at 57.4 °F (14.1 °C); the AquaBundance tank was 68.2 °F (20.1 °C). This differential shows how much heat the aquarium lost during the night. On the other hand, we had to cover the glass tank during the day to keep it from getting too hot.

Practical Aquaponics bathtub system.

Practical Aquaponics

Bathtubs

Guidelines — This can be a great way to make a fun, whimsical, 55- to 60-gallon (200-liter) aquaponic system out of inexpensive recycled materials that you know to be food safe. I'd like to create a pink one with claw feet some day. Murray Hallam has created a guide called *Bathtub Aquaponics* that takes you through each step of creating such a system.

Pros — What better use could there be for all those bathtubs in landfills?

Cons — This may not match the look of the rest of your backyard furniture! Also, again you are limited to a 55- to 60-gallon size.

Manufactured aquaponic kits and components

Guidelines — There are a few turnkey aquaponics kits on the market today, and given the rapidly growing awareness and corresponding popularity of aquaponic gardening, there will undoubtedly soon be many more. System kits can be a godsend if you want to garden aquaponically but aren't particularly handy and don't want to bother with collecting parts and doing it all yourself. The research and testing has, theoretically, been done for you. Kits are often also far more attractive and "finished" looking than DIY setups. Of course, there is a price for all this; you will probably pay more for a system kit than you would if you collected components and built the system yourself.

Since buying an aquaponic kit is potentially a large purchase in a new and fast-evolving industry, however, the buyer should beware. Here are a few guidelines to consider when evaluating aquaponic system kits:

• Manufacturer experience with aquaponics — If they haven't been growing aquaponically for at least a few years you should avoid buying a system from them. You don't want to be their guinea pig. This is not just a self-watering planter. It is not an aquarium with plants. It is not hydroponics, nor is it aquaculture. It is a fully integrated ecosystem with several living

AquaBundance system.

The Aquaponic Source, Inc.

elements. Part of what you are purchasing is the experience of your manufacturer. Make sure they have it.

- Manufacturer reputation — Ask for a list of current system owners, especially those whose systems are at least a year old. Contact them and find out how their experience with the vendor has been. Search for the vendor in the search engines and see what comes up. Check the better business bureau. Search for complaints on Twitter. Best of all, search the aquaponics forums and ask questions there. This is where folks are talking about aquaponics every day. They know who has a good reputation and who doesn't. Trust me. You will be glad you checked.

- Materials — The best materials for making aquaponics kits are polyethylene and fiberglass. Why? Because they are strong, they won't break down and corrode in a moist humid environment and they are non-toxic. Some of the early kit aquaponic manufacturers used wood for their kits because wood is easy to work with and easy to couple with supplies from the local big box hardware store. The problem is that wood breaks down very quickly in an outdoor environment unless it is constantly sealed and maintained. These kits often fell apart after less than a year of use.

- Also avoid any kit with copper or zinc components. While the plants can handle small amounts of these elements, and even require a trace amount, they are toxic to fish and if present, might leach into your system.

- Includes everything you need for a successful experience — Ask to see the user instructions or call the helpline. Make sure they take you through every step of system setup, cycling, planting and maintenance. The kit should also include a pump, timer or siphon, and testing and other supplies to get you through the first month or so of operation.

Pros — A high quality kit and/or components made specifically for aquaponics will save you a lot of time and heartache because the research and testing has already been done. You can immediately concentrate your time on enjoying aquaponics rather than designing and building, and possibly rebuilding, an aquaponics system. Also, these systems are generally much more attractive than anything you could pull together yourself.

Cons — Cost. While I know of no kit manufacturers who are getting rich selling aquaponics kits, they do need to make a living. You are paying a certain amount to their business, and a certain amount for the fact that

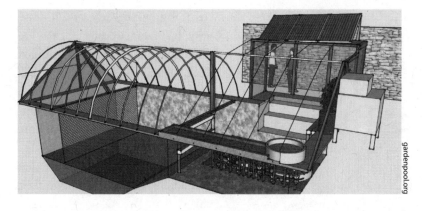

Garden pool design.

aquaponics is still new, so the volumes are low, making the cost of production high.

Swimming pools and ponds

While these can be an ideal basis for aquaponic systems, the challenge is trying to get to the right stocking density. The average swimming pool holds approximately 20,000 gallons (75,000 liters) of water. The ideal stocking density would be about 4,000 pounds (1800 kg) of fish, so you would need 20,000 / 7.5 = 2,667 cubic feet (75 cubic meters) of grow bed space! If your beds are 4 feet (120 cm) wide (the maximum width you can work comfortably on either side of the bed) you will have 167 linear feet of grow bed, or 21, 4′ x 8′ (120 cm x 240 cm) grow beds that you will need to keep planted and harvested. While this might be perfect for a commercial operation, it is too much for most hobbyists.

The way to get around this is to only use the deep end of your pool for the fish, and the rest for your beds. There is at least one family in Arizona who has done exactly that, and even put a hoop greenhouse over their pool system made of PVC and UV cloth for shade and storm control. They also raise chickens inside their "garden pool," and duckweed and black soldier flies to feed their fish. It is a terrific inspiration for anyone with a pool that has outlasted its usefulness! They can be found at GardenPool.org.

Vertical growing

Vertical growing is a technique where the plants grow in a vertical hanging structure rather than across a horizontal surface. There has been quite a buzz about vertical growing lately, and it's easy to see why. It is a fantastically efficient use of space, whether in an efficiently managed greenhouse or the window of a New York City apartment. If you are attracted to the notion of getting more growth out of a limited space you are going to want to explore vertical growing.

Here are some examples of ways that people are adding vertical growing to their aquaponic system:

- Window gardens — The windows of New York City are rapidly being filled up with recycled one-liter soda bottles growing plants. Simply cut the bottom cut off a bottle and hang it upside down, then fill it with grow media and hang it in succession from the top to the bottom of a window frame. Attach this to a thriving aquarium and you have a window aquaponics system that will fit in any apartment, no matter how small.
- PVC towers — 4-inch (10-cm) PVC filled with gravel or Hydroton® works well for vertical growing in an aquaponics system. You can either position it into your grow beds or add a screen to the bottom and hang the tower directly over your fish tank. To create growing spaces slit the pipe half way through the diameter, then use a hot air gun to soften the plastic above the slit. Once it is soft remove the heat and pushes it in with a screwdriver, leaving the screwdriver in place until the PVC cools.
- There are also ready-made retail versions of this idea now on the market called ZipGrow Towers. They are 4-inch (10-cm) rectilinear tubes that come in various lengths and come filled with a special media that feels a bit like

Pocket tower.

Christopher Muns

a Brillo pad. They are easy to plant, and based on our own greenhouse testing, work quite well.

- Stacking towers — there are a few brands of Styrofoam stacking towers on the market that can easily be adapted to aquaponics. These systems were designed as "drain to waste," however, so the key is to create a recirculating system to return the water to the fish tank.

Grow beds and fish tanks — conclusion

The grow bed(s) and fish tank(s) are the anchors for your aquaponics system, so you will want to select them carefully. They will define the footprint of your system and its ultimate usefulness and flexibility. They will also define your system aesthetics and therefore how you and others will experience it. Is it a bathtub system? An AquaBundance kit? Vertical aquaponics? An IBC tote system? Careful analysis now will get you off to a great start developing your aquaponic garden.

Aquaponic Grow Beds and Fish Tanks Rules of Thumb

Steps for planning your system
Determine the total grow bed area in square feet

ZipGrow tower.

- From grow bed area, determine the fish weight required (in pounds or kg) using the ratio rule 1 lb (500 g) of fish for every 1 sq ft (0.1 m²) of grow bed surface area, assuming the beds are at least 12 inches (30 cm) deep.
- Determine fish tank volume from the stocking density rule above (1 lb fish per 5–10 gallons of fish tank volume or 1 kg of mature fish per 40–80

liters). When your fish are young and small, reduce the number of plants in proportion to the size of the fish and their corresponding feed rate / waste production.

Grow bed and fish tank

- Start with a 1:1 ratio of grow bed volume to fish tank volume. You can increase that up to 2:1 once your system starts to mature (4–6 months) if you want to.
- Must be strong enough to withstand the lateral and downward forces of the media, water, and plant roots.
- Must be made of food-safe materials and should not alter the pH of your system.

Grow bed

- The industry standard is to be at least 12 inches (30 cm) deep to allow for growing the widest variety of plants and to provide complete filtration.
- Be sure to create or purchase a media guard to facilitate easy cleaning of your plumbing fittings.

Fish tank

- If you have flexibility here, 250-gallon (1000-liter) or larger seems to create the most stable aquaponics system. Larger volumes are better for beginners because they allow more room for error; things happen more slowly at larger volumes.
- You need at least a 50-gallon (200-liter) volume to raise a fish to 12 inches (30 cm) ("plate size").

8

Plumbing

"As soon as I learned about aquaponics, I became obsessed.
It just made so much sense to me. I couldn't stop researching
and ultimately built a backyard suburban aquaponics farm, which
provides a large quantity of organic fruits and veggies to my family
as well as those in my community through our buying club.
Learning to build a system from scratch using repurposed/recycled
materials whenever possible, combining different AP methods
and growing vertically (for space saving) was challenging,
but the rewards have been plentiful! There is still much
to learn and that is what makes it so interesting.
Every time I notice something else sprouting up or ready to
harvest, my heart leaps! Just the other day our seven-year-old
daughter came running in bursting with excitement,
as she had just picked the first ripe strawberry.
I'm thrilled to be part of the AP growing community!"

— Michelle Silva, Sarasota, Florida

Your pump and the piping it sends the water through comprise the circulatory system of your aquaponics ecosystem. Think of your pump as the heart, the pipes as the veins and arteries and the fish water as the blood. Like your own body, an aquaponics system will die quickly if it has a malfunctioning pump, clogged pipes or an erratic timing mechanism.

Why are we doing this?

In a flood and drain style aquaponic system, watering, or "flooding" the grow bed delivers nutrients to the plants and the bacteria. When the clean water flows back down to the fish tank, the draining action in the grow bed draws atmospheric oxygen deep through the media to the plant roots, the bacteria and the worms. This action mimics what takes place in a bog or a wetland. The action of flooding and draining delivers fertilizer to the surrounding plant life and helps clean the waste in the water.

The elements of your plumbing system

As I've already mentioned, you will need a pump to move your water, pipes and tubing for the water to move through and some mechanism for triggering the draining of your grow beds. Depending on the type of system you have selected (Basic Flood and Drain, CHOP, 2-Pump or Barrel-ponics®), the pump is either moving water from your fish tank into the grow beds (Basic Flood and Drain and 2-Pump), from the sump tank into the fish tank (CHOP and 2-Pump) or from the fish tank into the flood tank (Barrel-ponics®). If the water isn't flowing directly into the grow beds it arrives there through gravity (CHOP and Barrel-ponics®). Once the grow beds are full they drain, either when the pump is shut off (timer) or when a siphon or flush valve is triggered.

In this chapter I'll discuss considerations for pumps, pipes and draining mechanisms in a flood and drain aquaponics system.

The pump

Most aquaponic gardeners use submersible pond pumps to lift the tank water to their grow beds or from the sump tank back to the fish tank. These pumps are built to operate in a water environment with fish waste and plant debris. Magnetic drive (or mag-drive) pumps are best because the motor is in its own sealed compartment and should never leak any oil out into the fish tank water. When you select your pump, start with a trusted source of high-quality pumps. This is not the place to skimp! Read online reviews before locking in on a particular brand.

Pumps are generally described with two parameters: flow rate and head pressure.

Flow rate

When shopping for a pump, start with this simple rule of thumb — choose a high-quality pump that can, at a minimum, cycle the entire volume of your tank in an hour. If, for example, you have a 100-gallon (375-liter) tank, then you will want a pump that can pump at least 100 gallons (375 liters) to the height of the grow bed every hour. While your plants would probably be fine with less flooding, this insures that the fish are kept in clean, continuously filtered water.

Now, if you use a timer that is set to run for 15 minutes, then shut off for 45 minutes for that same 100-gallon tank, you will need to start with a pump that can pump a minimum of 400 (4 x 100) gallons per hour to the height of the grow bed. This is because you will only have a quarter of an hour to cycle through all that tank water.

Head pressure

Now, add to that how far the water needs to be "lifted" from the tank. This is also known as the "head height," with "head" being loosely defined as any resistance to the flow from the pump. The higher the water needs to go, the more oomph, or gallons per hour (gph) power, your pump will need to have. Pump descriptions typically include a chart showing how much the flow rate is degraded by increasing the head height. At maximum head height there is no longer any water coming out the top of the pipe. Also consider that over time, the inside of the pipes in your aquaponic plumbing will accumulate a buildup of solids. This is another form of water flow resistance that should be taken into account when sizing your pump.

Here is a sample chart for Active Aqua pumps showing how performance is affected by head height.

Pump head height chart.

Consider the AAPW1000 pump represented by the outermost line to the right. You can see that at two feet (60 cm) of head height water will flow at the full 1000 gph, and at about 13 feet (4 m) of vertical lift (the "head") no water is flowing. At seven feet (2 m) you will have about 600 gph. But again you also need to consider resistance caused by future solids buildup in your pipes and oversize your final pump specifications by 20 to 30 percent.

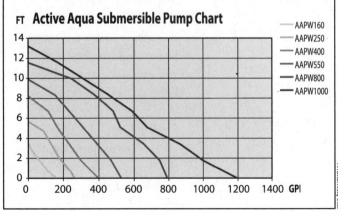

The pipes

Most piping used in aquaponics systems is PVC (polyvinyl chloride) or CPVC (chlorinated polyvinyl chloride) because they are food-safe, durable, rust-proof, widely available, cut easily and are inexpensive. Plus there is a huge range of fittings readily available, including tees, valves, corners, end caps, etc. Planning plumbing projects with PVC is frankly a lot of fun. My husband has decided that using PVC is like the grown-up version of playing with an Erector set. When we were first plumbing our greenhouse systems, he would eagerly pop out of bed in the morning so he could go outside and play with his toys.

PVC fittings can be threaded or slip fit. As the names indicate, threaded pipes screw together to connect, while slip fit pipes use sleeves that slip into one another. Pipe fittings are organized by male and female connectors as follows:

- Male-threaded: Threads are exterior, made to screw into the inside of a larger-diameter pipe end with internal threading.
- Female threaded: Threads are interior, made to receive male threaded pipe fittings.
- Male slip fit: No threads, made to slip into a slightly larger female sleeve.
- Female slip fit: No threads, made to receive a narrower male slip fit.

Pipe fittings are identified by pipe material, inner and outer diameter of the pipe and the type of fitting: threaded or slip, male or female. The ends of pipe fittings are slightly larger than the rest of the pipe to accommodate connections without narrowing the inner diameter (ID) of the pipe. This keeps flow consistent.

I recommend using standard, half-inch (12-mm) inner diameter PVC throughout your aquaponics system as it is large enough for adequate water flow without becoming clogged with solids, and because it is a standard size you won't have any problem finding it, or fittings for it. Larger sizes are also fine, but don't go any smaller. If you are going to use a larger size recognize that if you need to introduce some flexible hose into your system the larger the diameter of the flexible hose, the harder it is to find and the more it costs. In fact, we have been unable to find one-inch inner diameter flexible tubing that is opaque (clear tubing allows light through, which fosters algae growth).

I also recommend not gluing together any pieces of your aquaponics plumbing with PVC glue unless you absolutely have to. I'm less concerned with toxicity of the glue (there is no real concern here) than I am with ease of maintenance. You don't want anything standing in the way of quickly and easily taking your plumbing apart should it need a good cleaning.

Speaking of cleaning, the best way to clean your pipes is to use a hose with a high-pressure nozzle on it. I also have a chlorine filter on any hose that I use in my aquaponic systems so I never worry about killing the bacteria when I'm cleaning out the pipes or any other surface of my system that might harbor good bacteria. These inline dechlorinating hose filters are not expensive, and save you from ever worrying about chlorine.

PVC is not without its detractors. You may have heard of health and environmental concerns around the substance. I've evaluated the available reports on PVC and concluded that most of the healthy risk discussions are focused on softening agents, also known as "plasticizers" or "Phthalates," and their use in the toy industry. There is very little evidence that PVC or CPVC plumbing has any long-term health problems, except in high-heat environments. There are valid environmental concerns, however, both in the processes used to create PVC pipes and in their disposal. The best we can do here, I'm afraid, is to apply the adage "reduce, reuse, recycle" and try to find used material whenever possible.

While 66 percent of all home drinking water systems are made with PVC or CPVC, the second most widely used rigid piping for water distribution is copper. So, is copper a viable choice for aquaponics? No! You should avoid copper because it is absolutely toxic to fish, plus it is extremely expensive.

The alternative to using PVC is a rigid irrigation product called HDPE (high-density polyethylene). HDPE is also a petroleum-based product that doesn't break down in hot environments. It is #2 recyclable. The downside is that there aren't as many fitting choices nor are the fittings as elegantly designed for unobstructed water flow through HDPE as PVC fittings, and it is less widely available than PVC. An option is to use HDPE for your straight runs, then switch over to PVC for some of the more complex corners and connections.

Attaching pipes to the grow bed(s)

Whenever you drill a hole in your grow bed(s) in order to attach a water-in, water-out or overflow pipe, you create an opportunity for a leak. So before you drill, be sure you know how you will create a proper seal. There are many ways to do this, but the three that most aquaponic gardeners use are as follows:

Marine-grade silicone

UniSeal.

This is the least expensive, but also the least effective, method of sealing a pipe opening. While it will create an adhesive seal between PVC and most surfaces, it will not adequately adhere to polyethylene. It will also break down over time and need to be periodically reapplied. Finally, it creates a semi-permanent seal between the pipe and the surface of the grow bed so it can make maintenance challenging.

Uniseals®

These are black inserts made of a rubber-like material called DuPont Alcryn®. Uniseals® are designed to hold a pipe using pressure and are good to 40 psi. They come in a variety of sizes designed to fit a variety of PVC pipes. Your PVC pipe will slide through the center of the Uniseal®.

Bulkhead fitting

A bulkhead comes in two threaded pieces, includes a gasket and is hollow in the center so that water can pass through it. The first piece has male threads with a gasket on it. This piece comes in from the top (the inside of the bed) and passes through the hole in the bed. The other piece has female threads and acts like a nut. It screws on the male threads of the first piece from the underside (outside) of the grow bed. When you tighten the nut it compresses the gasket situated on the inner surface of the grow bed, creating a leak-proof seal. Bulkheads accommodate attaching separate PVC

Bulkhead fittings.

pipes from above and below your bed and come in both threaded and slip varieties so you will have a choice as to how you attach your pipes to your bulkhead fitting.

The timing mechanism

Core to the notion of flood and drain aquaponics is the periodic draining of the water back into the fish tank or sump tank that has been pumped up into the grow bed. As mentioned at the beginning of this chapter, the draining action pulls atmospheric oxygen through the grow bed media to the plant roots, bacteria and worms. Think of the timing mechanism that enables this as you would think of the signals that regulate your heartbeat. You can either go with a natural mechanism (an autosiphon) or a mechanical "pacemaker" device (a timer).

Mechanical and electronic timers

The easiest way to start your first aquaponics system is with a timer. When the timer is in its "on" cycle, it will send power to the pump for a specific amount of time. When the water reaches the desired height (typically 10 inches / 25 cm) the water exits the grow bed through an overflow pipe.

15-minute timer.

When the timer shuts off, the water will drain back through the pump or through a hole in the bottom of the grow bed that drains water at a slower rate than the water was coming into the grow bed.

The 24-hour timers that are most available and most reasonably priced are mechanical (they have moving parts) and work in 15-minute increments. There are a variety of electronic timers on the market but they tend to be expensive and most likely offer more capability than you will need. Given this, I discuss only mechanical timers in this section.

An aquaponic timing cycle typically runs for 15 minutes on and 45 minutes off. Experienced aquaponic gardeners have found that this schedule filters the water well, while allowing sufficient drying time for the plant roots. This also corresponds to the simple calculations of the pump flow rate we were discussing above. If you use the 15 on and 45 off schedule, you simply need to make sure that the pump flow rate (taking head height into

Hydrofarm, Inc.

account) is at least four times the number of gallons in your tank. This will ensure that you pump the entire water volume in your tank through your grow bed at least once every hour.

If you are coming to aquaponics from the hydroponics world, you have probably noticed that this is much more frequent flooding than a typical hydroponics flood and drain cycle of 15 minutes every four hours. This is because you need to filter the fish tank water, a problem that doesn't exist with a hydroponic system. The plants would probably be fine with a longer flood interval, but the fish may not.

I mentioned earlier that there are two ways of draining the grow bed: (1) back through the pump via the same hole the water enters the bed, or (2) through a separate exit hole that is smaller than the pipe the water is entering through. In the first configuration which is shown below, the water enters the grow bed through a hole in the bottom of the bed (A) and rises until it reaches the top of the overflow pipe (B). It stays at this level until the timer goes to the next cycle and turns the pump off. Then the water drains back through the pump (C) and the bed is emptied.

The primary benefit of this arrangement is that you will probably not experience any clogging issues. Since the water is leaving the bed through the same pipe it came in through, the force of the water being pumped into the bed clears out anything that would build up and block the water's exit. The downside of this setup, however, is that the solids are entering the bed in the same place the water is draining out of, so they tend to accumulate there.

F&D drain through pump.

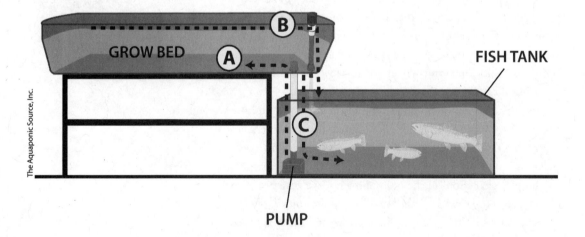

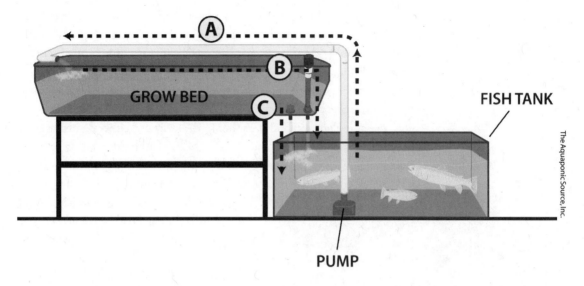

The Aquaponic Source, Inc.

Now consider the drawing above. Here the water comes in over the top of the bed (A), again rise to the top of the overflow pipe (B) and then drain out through a smaller opening in the bottom of the grow bed (C) or a small hole in the bottom of the overflow pipe.

Drain away from pump.

This is a little more complicated because you have to be sure that the water is always coming in at a faster rate than it is going out. The small holes can also get clogged with debris flowing in from the bottom of the grow bed.

The advantage of this type of grow bed plumbing is that you can send the water into your grow bed wherever it suits you. You can also add a water distribution grid if you like, for the most even distribution of solids across your bed. The disadvantage of this plan is that the solid waste is always draining down through a smaller opening, and without the upward force of the pressurized water moving into the bed, this opening can get clogged.

The other important player in grow bed plumbing is something called an overflow, or stand, pipe. This is what stops the water level from rising too high in your grow bed. This pipe has an opening that allows the water to flow out of the grow bed for faster than it is coming in. The overflow is as tall as you want the water level in your grow bed to be. Ten inches (25 cm) is optimal in a 12" (30-cm) -deep bed because it gives an inch or two of dry media at the top of the bed, plus it allows for a little extra height for the water to flow over the top of the overflow pipe.

The way that these pipes work in tandem begins when the timer goes on and the water is pumped into the grow bed through the inlet pipe. The water level rises until it reaches the overflow pipe. The water is held at a constant height of 10 to 11 inches (25 to 27.5 cm) inside the grow bed because once the water reaches the top of the overflow pipe, it immediately begins to flow through the overflow pipe and back down into the fish tank. After 15 minutes, the pump goes off and the water in the grow bed either drains through the inlet pipe and through the pump or, alternatively (depending on your chosen design), through small hole(s) in the bottom of the grow bed or at the base of the overflow pipe.

Any plumbing inside your grow bed should be contained within a larger perforated pipe or media blocker to block the media while still allowing water to flow freely in and out of the grow bed. Six-inch PVC with quarter-inch (7-mm) holes drilled on 2-inch (5-cm) centers works well for this. The top of each pipe should also be protected with a pipe screen to further prevent any media from entering and plugging the pipe.

The advantage of using a timer-based system is that it is easy to set up and operate. The disadvantage is that turning the pump on and off can shorten the life of the pump. Plus there is one more device that can fail — the timer — that doesn't exist in a non-mechanical siphoning system.

Timing without using a timer

There are two draining systems typically used in media-based aquaponics that have no inherent timing mechanisms and yet do time your system. These are autosiphons and flush valve systems. Autosiphons are more widely used so I will go into enough detail on these for you to build your own. Flush valves are largely used in

Media blocker.

The Aquaponic Source, Inc.

Pipe screen.

The Aquaponic Source, Inc.

the Barrel-ponics® systems so I will refer you to the Barrel-ponics® guide for detailed instructions.

Autosiphons

In the autosiphon (also known as a bell siphon) system, the pump is always on and water is constantly pumped from the fish tank into the grow bed. As the water rises, it fills the interior of the siphon positioned within the grow bed. When the water reaches the desired maximum height, it spills over a pipe within the siphon and creates a low-pressure area within the siphon that triggers the siphoning action. The siphon rapidly draws the water from the grow bed and into the fish tank until the grow bed is drained. At this point air enters the siphon, the low pressure within the siphon is lost and the siphoning (draining) action stops. Since the pump is always on, the grow bed begins to fill once again and the cycle repeats.

This may seem confusing at first, but is worth trying to understand because there are benefits to this approach. Because a bell siphon system doesn't wear on your pump by turning it on and off, it will extend the life of your pump. It also eliminates the need for a timer and creates a slightly better growing environment for your plants, worms and bacteria because the sucking action pulls the water from the grow beds so quickly that even more oxygen is drawn deep into the media. Finally, if you add an aeration bar for your fish tank on to your plumbing system (more on this soon), an autosiphon will also enable more air to be added to your fish tank.

Building an autosiphon

A siphon has three components that fit together in a nested configuration.

- The innermost piece is called the "stand pipe." The stand pipe is attached to the floor of the grow bed by screwing into a bulkhead fitting that goes through the grow bed (A) and exits to

Siphon stand pipe.

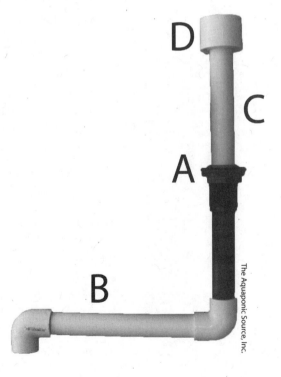

The Aquaponic Source, Inc.

a pipe that returns the water to the fish tank (or sump tank) (B). Above the connection to the bed floor is a PVC pipe (we use one-inch/25-mm) that is inserted into the bulkhead fitting (C). A funnel-shaped part is then attached to the top of the one-inch (25-mm) pipe (D). The purpose of this funnel piece is to create a "venturi" point where the water flow narrows and blocks the flow of air upward through the underside of the grow bed. This action enables the low-pressure area at the top of the funnel that triggers the siphoning action.

- The middle piece is called the "bell." It is a pipe long enough to fit over the stand pipe with an inch or two to spare. The top is sealed and air tight, usually with a pipe cap. The bottom of the pipe sits on the floor of the grow bed and has several notches about half to one inch (12 to 25 mm) above the bed floor to allow water from the grow bed to fill the bell as the water rises in the grow bed. As the water rises in the grow bed, the notches in the bell will be covered by the water. This will create an airtight seal within the bell. When the water rises to the top of the stand pipe, water will begin to spill into the funnel and down into the stand pipe. When there is sufficient flow into the stand pipe, the downward water flow will block the air from coming into the bell through the bottom of the stand pipe. This will facilitate the creation of the low-pressure area within the bell. When almost all of the water has drained from the grow bed, the water level will reach the notches in the bottom of the bell. At this point, air will enter the bell through the notches, the pressure in the bell will equalize with the atmospheric pressure and the siphoning action should stop.

- The outermost piece is the same "media blocker" we described in the Timer section above. Its purpose is to keep stray media from traveling down the pipes and clogging them.

Siphon bell.

The rate that water is coming into a grow bed that is being drained by a siphon should be regulated by a valve or a tap, because it needs to be appropriate for the chosen siphon configuration. This is because the flow rate needs to be great enough to trigger the siphoning action but not so great that the siphon will not stop when the water reaches the notches at the bottom of the bell. Siphons rarely work on the first try. They need to be adjusted by increasing or decreasing the flow rate. This can only be done if you have a valve on your inlet plumbing.

If there is not enough water flow, the siphon will not start. Use the valve to increase the flow to your grow bed until the siphon triggers reliably. If your pump is already at full volume, you have three choices:

- You can use a different, more powerful pump.
- You can decrease the width of the PVC inside of your stand pipe by inserting narrower piece of PVC as an inner sleeve. This will cause the air coming into the siphon from under the grow bed to be blocked more quickly and trigger the siphon at a lower flow rate.
- You can add a horizontal pipe extension to the outflow pipe that comes out from the underside of the grow bed (the outflow pipe is the pipe through which the water flows as it leaves the stand pipe and travels through the bulkhead in the floor of the grow bed). This horizontal pipe will add some "back pressure" to the system and thereby reduce the flow rate needed to trigger the desired siphoning action.

If there is too much water flow, the siphon will not shut off. You can fix this by simply turning the valve mechanism to slow down the flow rate into your grow bed.

To build a bell siphon you will need the following parts:

- 1-inch (25-mm) plastic bulkhead fitting with slip joints on both ends
- 3 lengths of 1-inch (25-mm) inner diameter (ID) PVC (lengths depend on desired height of siphon, and desired lengths of vertical and horizontal drain pipes)
- 2″ x 1″ (50 x 25 mm) PVC bushing
- 2 1-inch (25-mm) 90-degree PVC elbows
- 4-inch (10-cm) ID PVC pipe with half-inch (12-mm) holes drilled all around the bottom to let water flow in (or use machined slots for a nicer finish)
- 4-inch (10-cm) end cap (the end cap fits over the 4″ / 10-cm ID pipe to create the airtight bell)
- 6-inch (15-cm) PVC with holes drilled in to prevent gravel from getting into the siphon (or use machined slits for a nicer finish)

There are many excellent videos about the construction of bell siphons, and ready-made bell siphons are available for purchase on the internet.

Flush valve systems have been popularized by Travis Hughey and Barrel-ponics®. They work very much like a flush toilet. They require that you add a flood, or feeder, tank to your fish tank and grow bed setup. The flood tank, being similar to the tank of a toilet, is placed above the grow bed, while the grow bed, being similar to a toilet bowl, is placed below the flood tank, with the fish tank being the drainage point. Water is constantly pumped from the fish tank into the flood tank. A small siphon collects water, which then fills a container and acts as a weight. Once the water is heavy enough it triggers a standard toilet flush valve. This then allows water from the flood tank to drain into the grow beds and then into the fish tanks. The water-filled weight has a hole in it that makes it drain slower than it is being filled, so once the flood tank has emptied, the small weight begins to drain. Once the small

The evolution of the bell siphon by Affran

Every time a person sees one of my grow beds doing its routine flush they ask, "Is there a float valve controlling it?" When I say "No" they get this puzzled look on their face like they are trying to figure out how this flushing action could happen and what is controlling it.

I initially started aquaponics with a siphon flood and drain because I didn't realize that there were other ways to do it. It was very difficult because most of the information available on the internet suggested using a siphon for the water control and I really thought it was as simple as they said. This was also a blessing. If I had found other ways like deep-water culture, nutrient film technique, timer flood and drain or even the "breather tube siphon," I may not have experimented to make the siphon what it is now.

Using Bernoulli's Theorem, I realized that if we could provide additional assistance to the flow of the water through the stand pipe, then we could make a siphon more reliable. From my experiments, I found that the water in-flow required to start a siphon on a straight stand pipe is more than what is required to sustain water out-flow at the end of siphon. This creates a phenomenon that we call "equilibrium flow," where the siphon refuses to stop.

Equilibrium flow is a siphon killer. It would be impossible to use if it happened on your siphon. Various designs have been used to eliminate equilibrium flow. One of the designs that works is the use of a "breather tube" designed to suck air at a cut-off level, thus helping to stop it. Under an ideal scenario, this method works. However with fish solids, gravel, dirt and the weakening of your pump over time, it will fail and therefore require constant maintenance.

☞

weight has drained, it again is light enough to close the standard toilet valve, reinitiating the cycle.

Some other thoughts about circulating water

Irrigation grid for larger beds

If you have a grow bed with a planting surface greater than eight square feet (0.75 m²), you may want to distribute the water coming in from the tank through a device called an irrigation, or distribution, grid. Using PVC and various connectors and end caps you can construct a grid of pipes, all capped at the ends. Drill holes on the undersides of the pipe to allow the fish tank water to enter the grow bed at many locations. Start with the fewest number of holes you think you will need — it is easier to drill more holes than fewer.

In aquaponics we cannot afford to have the siphon fail. Thus, it is crucial that the siphon requires minimal maintenance, has no moving parts and can tolerate a wide range of water inflow rates to withstand a weakening pump. As I experimented with various designs, I constantly reminded myself of these criteria.

Let's face it. If you need to constantly check on a device, it will likely fail before you get very far. Having this in mind, I tried to design something that works without much maintenance. To overcome equilibrium flow, I needed to make it operate with less starting water and at the same time the outflow water needed to exert enough force to make it useable with a wide range of water in-flow.

Applying Bernoulli's Theorem in the form of a funnel in the stand pipe overcame a lot of the initial problems I encountered with a siphon. This additional "push" created by water flow caused a lower initial starting water flow and also caused the siphon to be able to tolerate a broader range of in-flow.

Slight back pressure is needed to assist siphoning, however too much will cause equilibrium flow. If the outlet tube is long enough, just the tube itself will create sufficient back pressure.

In situations where the outlet cannot be made long enough, elbow joints or a reduction fitting can be used to create this back pressure. This approach is widely used in many installations.

A lot has been done to this simple device to make it reliable. I hope this will help promote aquaponics to a broader audience and new enthusiasts won't be intimidated by the complexity of the physics that enables this seemingly simple device to operate.

— Affran's Aquaponics, AquaponicsMalaya.com.

Also, do not use a drill bit that is less than $^3/_{16}''$ (4.75 mm) because smaller holes will be prone to clogging with solid waste.

Index valve.

Indexing, or sequencing, valves

An indexing valve is an interesting addition to your plumbing if you are moving beyond a 1:1 grow bed to fish tank ratio, and in particular, toward a 2:1 ratio. In this case, you may need to adjust your design to avoid draining your fish tank by flooding all the beds at once. Essentially an indexing valve works by interrupting the water flow into the valve and then reinitiating it. Every time the water flow stops and then restarts, a new outlet on the indexing valve opens. With this feature, you can direct water in sequence to each of your beds. In this way, the water demand on your fish tank will only equal the amount of water required to flood one of your beds instead of all of them if you were to flood all of your beds simultaneously. These valves come in a variety of sizes to match inner diameter PVC pipes of 1″, 1¼″ and 1½″ and with a choice of four, six or eight outlets.

Aeration pipe or spray bar

I always add an aeration pipe, or spray bar, of some sort to all of my fish tanks to add extra oxygen into my fish tanks. This is a pipe that diverts some of the water being pumped into the grow bed straight back into the fish tank. Just like with the irrigation grid described above, an aeration pipe connects to the water

Aerating spray bar.

TCLynx, Aquaponic Lynx LLC www.aquaponiclynx.com

Luvandpeas, Aquaponics AKA Dave and Jenny Noorlander

going up into the grow bed through a tee valve into another pipe that has holes drilled across the underside and is capped at the end. The reasons for setting up an aeration pipe are twofold: you can never have too much oxygen in your fish tank and it becomes a simple redundant oxygenation system in case your other aeration device fails.

Backup pumps and aeration

It is important to have backup plumbing for your aquaponic systems. If your pump fails, your plants will be fine for a day or two but a few hours without moving water could lead to dead fish due to both oxygen deprivation and lack of filtration. I highly recommend that you have a second, backup pump available just in case your primary pump fails. And if you are relying on your plumbing system to supply most of the oxygen to your fish tank, you should also consider a backup aeration system, and what you might do if (when) your power goes out. Having some battery-powered aerators on hand might be a very good idea.

Aquaponic Plumbing Rules of Thumb

- You should flood and then drain your media-based grow beds. The draining action pulls oxygen through the grow beds. The least complicated way to achieve a reliable flood and drain system is to use a timer. While more complex, siphons are also excellent options for aquaponics.
- If you are operating your system with a timer you should run it for 15 minutes on and 45 minutes off.
- You should flow the entire volume of your fish tank through your grow beds every hour, if possible. Therefore, if you are running your pump for 15 minutes every hour, and you have a 100-gallon (375-liter) tank, you need at least a 400-gallon per hour (gph) (1,500-liter per hour) pump. Also consider the "lift" or how far against gravity you need to move that water and use the sliding scale on the pump packaging to see how much more power you need beyond the 400 gph.

9

Grow media

*"I'm in… All in… Maybe it's the satisfaction of knowing that
we can be self-sufficient? …. Not that we have to be … but
that we can be… if we need to… and we're walking around
with that in our back pocket …. Today, my siphons clicked off and
I got my plants in the gravel…. This evening with a drink in my
hand … listening to the water peel off into my sump … a sense of
self-reliance came over me … an odd feeling … one that I haven't
exactly put my finger on yet …. But I know one thing… I can
understand now why you guys name your fish…"*

— Darryl Hinson, Newllano, Louisiana

In traditional gardening the growing media, or soil, IS the garden. It is a deep, mysterious, black box from which life springs and later decays. It is the essence of our connection to the globe. "Earth" is simultaneously a planetary concept and a speck of dirt. We plant in the soil, water the soil and our plants grow and thrive according to the properties contained within the soil.

One of the most mind-bending aspects of aquaponic gardening for a traditional gardener is the manner in which soil is deconstructed. Rather than the black box that is soil in the earth, in aquaponics each property of the soil is carefully constructed in the media environment.

The media in an aquaponics system serves the following "soil-like" purposes:

Structural

The 12 inches (30 cm) of media in your grow beds provide a foundation and structure for the roots of your plants to embrace and provides a counterbalance for your plants in the case of high winds, or just gravity.

Surface area for the nitrifying bacteria — The bacteria in your aquaponics system require a dark, moist, nutrient rich environment in which to set up shop. The more surface area they have in which to become established, the more robust your biofilter will become. The ratio of surface area of a media to the media's volume is called the specific surface area (SSA) of the media, and this is stated as m^2/m^3 (or surface area per volume). A ball of expanded clay (lightweight expanded clay aggregate or LECA) or a volcanic stone will have far more SSA, and therefore bacteria, than polished rock of a similar size.

Biofilter for the solid waste

In media-based aquaponics, the solid waste from the fish is pumped into the grow beds along with the liquid waste. This is then broken down by heterotrophic, aerobic bacteria as well as the composting worms you add to your bed. Your plants will appreciate and thrive on the additional nutrients.

Home for worms

An aquaponic media bed that your system floods and drains with fish waste becomes an ideal environment for a thriving composting worm population.

Air and water exchange

The porosity of your grow media and the spaces among the media hold air and water between flood cycles that are vital to the health of your plants.

Soil vs. aquaponic media

Let's strip soil down to constituents that feed plants. There are the macro- and micronutrients, there are the microorganisms that break down dead organic stuff and make it available to plants, and then there is the water needed for a wet root-soil interface through which the DISSOLVED nutrients can then be absorbed by plants. FACT: Plants cannot take anything out of soil without this water interface. FACT: Soil by itself is merely weathered rock, which does not contain all the nutrients itself, but traps the resulting matter produced by the decomposition of organic substances. FACT: If you sterilize soil, only those minerals released by the decomposition of the parent rock will be available to plants. Thus UBERFACT: Soil is an anchoring medium to plants that may or may not, over time, release some of the stuff plants need to grow. In the greater scheme of thing soil, rock, charcoal does not matter — it is all the same thing for a plant as it plays the same role.

— Kobus Jooste, Aquaponic System Development, South Africa

Temperature moderation

The media acts as a blanket around the roots of your plants, protecting them from rapid fluctuations in temperature.

What is the best medium?

Most people use either gravel or expanded clay (Hydroton®) for their media and both are widely available. Expanded shale is a media type that is quickly gaining popularity. It has many of the wonderful properties of expanded clay, including being highly porous and lightweight, but is domestically produced and less expensive. In addition, there may be other candidates, depending on availability in your area (see the table below).

Expanded shale.

Here are some "must-haves" and "nice-to-haves" for you to consider when you select your aquaponic media.

Media must-haves

Must not change the pH of your water

Both initially and over time, the media must not give off anything that changes the pH of the system or contributes any nutrients to the system. Watch out for limestone and marble because they tend to create high pH environments due to their calcium carbonate levels. Diotamite, Maidenwell and Higrozyme all tend to drive pH down over time. Most river stones and lava rocks are inert and pH neutral.

Must never decompose

Your media should never break down or decompose. If it does, you may experience uncontrollable fluctuations in your pH and nutrient levels and a mess to boot. The decomposition process may also leach tannins into your water that could turn it dark and make it hard for you to see your fish. The risk of decomposition means you should not use soil, peat moss, wood chips or coconut coir.

Must be the proper size

Just like Goldilocks, you need to find media that is not too small, and not too big, but is just right. Media that is too small — for example sand, pearlite and vermiculite — will quickly become clogged with solid waste, too compact and not allow good air and water circulation around the root zones of your plants. Media that is too big — for example, large lava rocks — will create large air pockets where the plant roots won't comfortably establish themselves. The "just right" size for aquaponic media is about half to three-quarters of an inch (12–18 mm) in diameter.

Media nice-to-haves

Porosity

The more surface area you give bacteria to establish itself, the more robust and productive your system will be. Plus porous material holds air and water better than non-porous material, and weighs less. Properly sized lava rock is a great example of this.

Be easy to handle

Hydroton®.

Sharp edges are tough on plant roots and a gardener's hands. If you fill your bed with a rounded or smooth-surfaced media such as river rock or manufactured, expanded clay balls such as Hydroton®, you will thank yourself again and again.

No matter what media you choose, it must also fit within your budget. Because flood and drain aquaponics grow beds are best at 12-inches (30 cm) deep, the required media can be a real budget buster if you aren't careful.

The comparison chart below helps to explain the tradeoffs between the most popular forms of media used in aquaponic systems.

A quick test for gravel

I have heard several heartbreaking stories of aquaponic gardeners that set up huge systems using gravel, only to find out that the reason

	Expanded shale	Expanded clay (Hydroton)	River stone	Crushed stone	Synthetic
Weight	Three quarters the weight of stone	Half the weight of stone	Heavy	Heavy	Lightest — tends to float
Enviromental	Mined from a quarry	Mined from a quarry	Mined near local rivers. Larger environmental impact than engineered quarries	Mined from quarry or consists of crushed river stone	Made from petroleum
Origin	United States	Typically imported from Germany or China	Local quarry	Local quarry	Typically imported from China
ph Neutral (inert)	Yes	Yes	Take a chance. If the stone has any limestone, it will continue to raise the pH	Same problem as river stone	Yes
Easy on the hands?	Yes — even though the shale has been crushed, the kiln process rounds over all the edges	Yes	Yes	Typically very sharp and hard to dig in with bare hands	Yes
Expense (1=cheapest - 5=most expensive)	3	4	2	1	5

Media comparison chart.

they could not control their system's pH was because the gravel they chose had limestone in it. Don't let this happen to you!

First, if you can find gravel made from granite or other quartz-type rocks, or lava, which is sometimes referred to as scoria, you should be fine. These are all known to be pH neutral.

Otherwise, even if your local quarry has assured you that their gravel is pH neutral and contains no limestone, my advice is to run a quick test before you have a truckload delivered to your driveway.

Start by getting a small sample of the gravel you are considering using. Just a handful works well. Rinse it and put it in a cup of vinegar. If it fizzes,

The Aquaponic Source, Inc.

Granite gravel.

you don't want to use it. You can also just take some and put it in a cup of distilled water for a week or so and see if the pH rises from 7.0. If it does, you should look for another source of media!

Washing the media

No matter which media you choose, you should be prepared to wash it before starting your system. LECA will have a thin layer of red clay dust over the surface that will cloud your fish tank water and eventually settle to the bottom of your tank. Gravel and volcanic rock have probably been stored outside in a stone yard and are likely covered with a variety of dirt and waste products that your fish would rather not live with.

I've used several techniques for washing media. If you are able to work outside with a hose, and it is easy to maneuver, you might just want to use your grow bed. Set it outside on a raised surface so the water can drain out the bottom. Be sure to plug the holes with either screen fittings or a piece of screen door mesh material. I've also used a compost-sifting screen covered in screen door material.

If you need to actually transport the media indoors to wash it in a sink, I recommend using a camping dunk bag. I've found that these are inexpensive and will hold about eight liters of Hydroton® per bag. If you can convince someone to help you, get two dunk bags so one of you can fill while the other person washes. At twice the speed it is worth the extra cost!

Aquaponic Media Rules of Thumb

Any media you select

- Must be inert — i.e., won't alter the pH of the system
- Must not decompose
- Must be the proper size (½″–¾″ / 12–18 mm aggregate is optimal)

- The most widely used media types are LECA (lightweight expanded clay aggregate, aka Hydroton®), lava rock, expanded shale and gravel.
- If you choose gravel, understand its source and avoid limestone and marble as they could affect your pH.

Andrea's aquaponics story

I first heard about aquaponics while reading a blog about living a simpler life by a woman who makes her own soap and hangs her laundry on a line. My life isn't simple, and her extreme simplicity isn't for me — I just enjoy reading about her life since it's so different from mine. I live in Southern California, drive a sports car, and am a computer programmer. Oh, and I have a black thumb. On the other hand, I am (finally!) learning to cook and had a 30-gallon fish aquarium in an unused corner near the kitchen. I had been just patiently waiting for the last fish to die of old age so I could get rid of it. But now I thought I'd modify that into a small aquaponic system so I could grow some herbs for cooking.

That was about a year ago. The herbs grow faster than I can use them and I share them with friends and coworkers. It was so simple that I'm afraid I got a little overly enthusiastic about it. I now have a 10' x 20' greenhouse in my driveway next to the garage. It encloses a 340-gallon fiberglass fish tank (currently full of small tilapia) and a 100-gallon grow bed. I'm still building more grow beds to attach and am expecting to have 600 gallons of grow bed by the end of the year — plus some vertical towers. It is my sincere hope that by the end of the year I'll be growing the majority of my own produce and selling the excess to fund the purchase of any produce I don't grow myself.

The setup of the system has been quite an adventure. And I've sometimes found that my efforts to do something myself to save money have ended up costing more than buying a quality component would have. But in the end, I'm very happy with what I have built. And since I really only need to feed the fish daily, and test the water weekly, it fits in perfectly with my busy life.

But I admit I spend more time out there than is necessary for the required basic maintenance. The sound of the water is like a fountain and the warm humidity of the greenhouse is soothing. I love to just sit out there and listen to it and relax after a long day at work. And the plants shoot up so quickly I go out to gloat over them.

It hasn't all been perfect. I've killed some fish. I've repaired some leaks. I've even done a bit of ranting, raving, and swearing… but not much. It's been an amazingly fun journey. And out on the side of my house, on a slab of concrete, in the middle of the suburban California desert surrounded by bedroom communities and rat-race-focused people (like me), there's a little oasis of fish and plants where tomorrow's dinner is peacefully waiting to be gathered up.

— Andrea Keene, Yucaipa, California

10

Water

*"Aquaponics married my love of tropical fish keeping with my love
of gardening. It means much less work and more enjoyment for me,
including fresh veggies in the winter!"*

— Paul Letby, Winnipeg, Manitoba, Canada

Just as the media is the home you construct for your plants in an aquaponics system, the water is the home you construct for your fish. And just like with us, if the fish live in a home environment that is clean, kept at a comfortable temperature, free of toxins and with plenty of air to breath, they will thrive. But if any of these elements is missing or sub-optimized they will be stressed, stop eating, not thrive and possibly not survive. Said another way, water is the "lifeblood" of your system, so strive to get it right.

In this chapter, I'll go over all of the water environment factors important for your fish. You will learn how to manage water purity, temperature, oxygenation and pH.

Purity

The water for your fish environment should enter your aquaponics system as close to pure as possible. Ideally, it will be pH neutral and not contain any chlorine or chloramines. However, in the real world this usually isn't what you will find. If your water hasn't just come out of a water purifying system, you can assume that there is "stuff" in it. The "stuff" may or may not be harmful to your fish or another living component of your aquaponic system. Given this, you may need to do some adjusting to your water, but first you must understand what you are dealing with.

Municipal water supply

If your water is supplied through your municipal water district, you can probably find out lots of information about its overall quality, pH and additives such as chlorine and chloramines through the city website or through a simple phone call to their office.

Since municipalities tend to worry about acid corrosion of household pipes, water will typically come out of most home taps above pH 7 (neutral). It will also almost certainly contain chlorine, but that is easily remedied. To remove the chlorine from the water, just fill up your tank, turn on your pump and run the water through your system for a day or two. While the chlorine will naturally leave water in a few days, adding air to the process by running your system will get rid of the chlorine faster by accelerating the "off-gassing" process. During this process, the pH may change, so don't bother trying to modify it until you have first off-gassed your chlorine.

If you are just adding a small amount of water to your tank to top it up (e.g., less than 10 percent of the tank volume), don't worry about off-gassing the chlorine. This is too small an amount to worry about. However, you should be prepared to add larger amounts of water to your system in case of an emergency or for any other reason. I recommend one of the following. You can fill a separate water holding tank so that you always have a supply of dechlorinated water ready to use (as I just mentioned above, the water will lose its chlorine in a few days just from sitting in the tank). Another good choice is to buy a dechlorinating filter and install it in-line on your hose spigot or faucet. With this filter in place, you can fill your tanks directly from your tap knowing the chlorine is being removed by the filter. No waiting involved!

Chloramine is a much bigger problem than chlorine, and is more difficult to get rid of. Although sunlight (UV) eventually breaks down chloramine, it does not off-gas. If your municipality treats its water with chloramine, you should filter it from your water using activated carbon or break it down using a UV filter before filling your fish tank or find an alternative water source for your system.

Rainwater and well water

Rainwater catchment is an excellent water source for your aquaponics system because it is relatively pure and will be free of both chlorine and chloramines. By roofing over the fish tank, you can take advantage of a large surface area to direct rainfall straight into the fish tank, or a holding tank, depending on the current level of the tank water. Unfortunately rainwater is acidic in some areas, so if you are using it for your aquaponics system you might want to consider having a sample analyzed for acid and other impurities.

Wells are another good water source for your system. Again, consider having it analyzed, as most well water contains a variety of minerals that may, or may not, be good for your aquaponics system. Also, according to Dr. Lennard, "Many wells have water with high carbon dioxide content, so a good approach is to aerate well water for a couple of days before using it in your system to 'off-gas' the carbon dioxide. I have seen entire tanks of fish lost by adding carbon dioxide-rich well water to the tank."

Temperature

The type of fish in an aquaponic system dictates the optimal temperature at which the water should be kept. I go into this in some detail in the chapter on Fish, but note that all fish will have both an acceptable temperature range and a narrower, optimal temperature range. The broader, acceptable range specifies the high and low temperatures the fish can tolerate without being harmed or even killed. The narrower, optimal range specifies the temperatures in which the fish will thrive, grow the fastest, be the least susceptible to disease and, most importantly for aquaponics, will eat and metabolize the most food and create the most "waste."

Your best strategy is to choose fish species that match as closely as possible the climate in which you live. Why? First, heating and/or cooling the water in your system can be quite expensive, depending on how much you are trying to change the water temperature, your method for heating or cooling, how much you can insulate your water tanks and your cost of energy. Second, if your system should fail, perhaps due to a power failure, and the water temperature leaves the acceptable range, you may lose your fish. So, if you live in a cold climate, consider growing fish such as trout that thrive in cold water, and if you live in a more tropical climate, consider a warm-water species such as tilapia.

Unfortunately there are two problems with this simplistic idea. The first is that very few of us live in a climate that has a steady, year-round temperature. The second is that the plants, bacteria and worms need to be considered in an aquaponic ecosystem as well. They typically prefer warmer water to cooler water, although there are exceptions like spring crops that typically do better in cooler water.

Because the overlap between acceptable water temperature for fish and the acceptable water temperature for plants, bacteria and worms is greater with warmer-water species of fish than colder ones, it is more likely that you will want to heat your water than cool it. This is actually good news because there are far fewer options for cooling water than heating it, and the ones that are available are very expensive and energy intensive.

I recommend a dual approach to heating your water. First, attract and retain the heat that is naturally in your environment. Second, be prepared to increase the heat from there, if necessary.

Attract and retain heat

The most energy efficient way to heat your home is to use insulation to prevent the heat you have from leaving. Similarly, any strategy for managing the temperature of your fish tank water should begin with attracting heat and then retaining the heat you have.

The primary way to attract heat into your fish tank is to start with a black or other dark-colored tank. Dark colors absorb heat. If you already have a tank in mind that isn't black, or you want the option of not attracting heat when warmer temperatures arrive, then a dark insulating blanket or black plastic draping will work as well.

To retain heat, you need to think through all the possible escape routes for the heat in your fish tank, starting with the tank itself. Thick or double-walled materials will hold heat better than thin-walled materials.

If there is a chance that cold could seep through the floor of your tank, you should position it over insulating materials when you first install your system. Likely trouble spots include greenhouses with exterior walls that are not insulated down to the frost line, garages and basement floors. Also consider wrapping the sides of your tank in a thermal blanket and covering the top of the tank to minimize heat escape from the water into the air. Be careful, however, not to cover the tank so thoroughly that air exchange is impaired. Use mesh materials, wood slats or other breathable covers.

Adding heat

If your tank is small — 200 gallons (750 l) or less — you will probably be best off with a simple aquarium heater. The rule of thumb here is that it takes about one watt of power to heat a liter of water. So if you have a 200-liter tank you should have a 200-watt heater. In imperial measures, this comes out to the recommendations in the chart (right).

Aquarium heater.

Tank Size (gallons)	Tank Size (liters)	Heater Wattage
5–14	50	50 W
14–26	100	100 W
23–40	150	150 W
26–53	200	200 W
53–80	300	300 W
80–105	400	400 W

Aquarium heater chart

I recommend that you double up your heaters so if one stops working, you will still have the other giving you some amount of heat. So if you have a 200-gallon tank, two 400-W heaters should be sufficient. Or better yet, consider four 200-W heaters. If you have that many heaters, though, you might also want to invest in a separate temperature controller so they are all being driven off the same control device.

While this seems like a lot of power, it isn't quite as bad as it first seems. These heaters generally operate on thermostats so they will only turn on if the water temperature goes below the thermostat setting. And be cautious when they are on — these heaters become very hot, very quickly.

Here are some other water heating ideas:

• Swimming pool heaters for large fish tanks. One with a titanium heating element that won't corrode is most appropriate to a fish tank.

• Hydronic, radiant or "in-floor" heating of your fish tank, or your entire greenhouse, by running heated tubes through the floor. There are a variety of ways to heat the liquid that runs through the tubes, from solar to geothermal to a boiler running on natural gas, wood pellets or other biomass sources (corn cobs, cherry pits, etc.) to a "rocket stove."

Is it possible for the water to get too hot even for tropical fish? Yes, but the main problem here isn't the heat as much as a lack of oxygen. Read on.

Dissolved oxygen

Now you have water in your fish tank. It is relatively pure and at the right temperature for your fish. The next step is to infuse it with oxygen so the fish can breathe. And yes, oxygen is also critical for the health of the roots of your plants, the worms and the bacteria, so the more oxygen in the water the better off all the living elements of your aquaponics system will be.

Oxygen gas in liquid is described as "dissolved oxygen" or DO, and is measured in parts per million or ppm.

The maximum amount of oxygen you can dissolve into a liquid is called the "saturation point." Saturation depends on a variety of factors, including:

• Temperature — Cooler water holds more oxygen than warmer water.
• Altitude and barometric pressure — Oxygen is more easily dissolved into water at low altitudes than at high altitudes, because of atmospheric pressure is higher at lower altitudes.
• Salinity — As the amount of salt in a body of water increases, the amount of dissolved oxygen decreases.

Other factors to consider in estimating how much DO is in the water are how much biological activity is occurring in the water and how much movement there is (including bubbles breaking) on the surface of the water. Biological activity, such as decomposing plant or animal life, requires oxygen. This is one of the reasons why an algae bloom, or "green water," in a fish tank can have devastating consequences. When the fast-growing algae starts to die off it consumes oxygen as it decomposes. Plus, according to Dr. Lennard, large algal blooms can and do remove appreciable amounts of oxygen even when they are living. This occurs at night, when all plants have the ability to reverse their gas exchange approach and uptake oxygen and release carbon dioxide. Algae can therefore deplete oxygen overnight to a critical point.

Moving water will have the opposite effect, adding oxygen to your water. The more you disturb the water surface of your fish tank, the more oxygen you will be adding to your water.

Fish varieties have specific oxygen requirements that stem from their native environments. Because some tilapia come from African lakes that often suffer from poor water quality, they have evolved to be extremely tolerant of relatively low oxygen levels. Brook trout, on the other hand, evolved in the clear snow melt runoff found in fast-running mountain streams, so they require relatively high levels of oxygen.

When in doubt about the oxygen requirements of the fish you are raising, always lean toward more oxygen rather than less. The bottom line is, it is difficult to have too much oxygen in your fish tank. You will most likely hit the limitations of the saturation point long before that occurs. As Murray Hallam of Practical Aquaponics is fond of saying, "The only way you can have too much oxygen in your fish tank is if you are literally blowing your fish out of the water!" While this may be a slight exaggeration — you can actually cause something called "air bubble disease" in fish if you supersaturate the water by directly injecting oxygen — suffice to say that in most home situations, the more air the better.

In general, you should strive to keep your oxygen level in your fish tank above 6 ppm. Most fish will become stressed at a DO of 3 ppm, and many will die at 2 ppm or below. How do you know how much DO you have in your system? One way is to use an oxygen meter. These are fairly expensive, however. There are inexpensive test strips available, but I've found that these are highly inaccurate and not even worth the small amount of money they cost.

I recommend a simpler, two-part method for evaluating the oxygen in your fish tank. First, follow the three steps below to force as much oxygen as possible into your fish tank water. Second, observe your fish. If they are eating well and not gasping for air at the surface of your tank water, you have enough oxygen.

Adding oxygen

Oxygen is added to your water when the surface is broken and an exchange of atmospheric oxygen with the water takes place. There are three primary ways to break the surface water and drive oxygen into an aquaponics fish tank:

1. As water drains from the grow beds and returns to the fish tank. There is an opportunity for oxygenation to occur when water falls back into the fish tank from either the grow beds or the sump tank. The higher the water source as the water falls into the fish tank, the more splashing will occur, and the more oxygen will be driven into the tank.

2. If you are using a timer-based system, you can also take advantage of the return water splash during the overflow in the flooding cycle. Just be sure the overflow water enters the fish tank above the surface rather than using a pipe to send it below the water surface inside the tank.

3. Water diversion. What if rather than sending all the water flow from your fish tank into your grow beds, you diverted part of it back into your fish tank? You can either send it straight back into the fish tank in a single stream or attach an aeration spray bar and have as many points of entry into the water as possible. Just imagine how much you could disturb the surface of your fish tank if you used a siphon and your pump was always on. This is also a good way of relieving back pressure on the pump.

Supplemental oxygenation

Rather than adding oxygen by splashing water into your tank, you can create bubbles within your tank by using an aeration device. These send air through a narrow tube connected to a valve on the aerator at one end, and an air stone or other diffusion device on the other. When this is powered, the effect is to send small bubbles through the water in your tank that break the surface of the water when they arrive at the top. While a small amount of air is added to the water as the bubbles travel upward, most of the benefit happens, again, when the surface of the water is broken.

Aerating spray bar.

I recommend that you always have a source of supplemental oxygenation for your fish tank as a backup in case your water pump stops functioning. You might also be well served to seek out an AC/DC aerator that will switch to battery power if your AC power is interrupted.

pH

The last important consideration for the water environment is pH. Depending on how recently you studied chemistry, you may recall that pH stands for "power of hydrogen" and is defined as a negative logarithm of the hydrogen ion activity in a solution. An easier to understand notion is that it measures the acidity or basicity of a solution on a scale of 0 to 14, with a pH of 7 being neutral. Pure water is said to have a pH of 7, while a pH of less than 7 is acidic and greater than 7 is basic. The scale is logarithmic, which means that each unit change equals a 10-times increase or decrease of acidity. Therefore 6 is 10 times as acidic as 7, and 5 is 100 times more acidic. Small changes in pH actually result in large changes in acidity.

Just like with temperature, most living things have a very specific pH range in which they can survive, and an even narrower range in which they thrive. In aquaponics, there are four living things co-habiting the same ecosystem — your fish, your plants, the worms and the bacteria — so you want the pH of your water in a range that is compatible with all four. For fish, this is optimally around 6.5 to 8.0; for plants, around 5.0 to 7.0; for red worms and bacteria, 6.0 to 8.0. Thus, the optimal pH is a compromise of around 6.8 to 7.0. This is okay for the fish (as it protects against ammonia toxicity) and the plants (as it allows them to take up the nutrients they require for growth) and is good for the red worms and bacteria.

I recommend testing pH at least weekly, and as frequently as three to four times a week, using your API Freshwater Master Test Kit or a digital pH meter. During initial cycling, pH will tend to rise. This is because the water coming from your tap is most likely above 7.0, and there are no natural processes in place during cycling to pull the pH down from there.

After cycling, pH will probably regularly drop as a natural result of the biological activity in your system. The living things (including the plant roots) are all taking up oxygen and excreting carbon dioxide during respiration and photosynthesis. However, according to Dr. Lennard, the pH drop is mostly associated with the nitrification of ammonia to nitrate, which produces a lot of hydrogen ions, which makes the water acid. This accounts for about 90 percent of the pH drop in a system.

This will require adjusting the pH with a buffering agent. In non-technical terms, this means adding an agent that will move your pH toward neutral and keep it there by resisting the forces that are pulling pH downward. You can test every day and adjust with a base or an acid, but if the system is not buffered, the agent causing the swing will return and you will be cursing again.

You should take action to adjust the pH in your system if it drops below 6.4. The best method for raising and buffering pH is to alternatively use calcium hydroxide (hydrated lime) or calcium carbonate (agricultural lime) with potassium carbonate or bicarbonate (found in beer brewing supply stores) or potassium hydroxide ("pearlash" or "potash" — handle carefully as this can be extremely corrosive and burn skin easily). These also add calcium and potassium. While supplementation is normally not needed in an aquaponics system, your plants will appreciate it nonetheless.

While they work, be cautious about using natural calcium carbonate products such as eggshells, snail shells and seashells. They don't do any harm, but they take a long time to dissolve, release the carbonate and affect the pH. So a likely scenario is that you add one of these items, check pH a few hours later, find nothing has changed, so you add more. Then after a few days, the pH suddenly spikes because you have added so much. The best way to solve this problem is to make sure that the additive is easily removable. You could,

What is a buffer solution and why should you care?

A buffer solution changes pH very little when an acid or base is added to the system — up to a point — at which time the buffer "breaks," and the pH changes quickly and substantially (something you would not want to have happen). You should care because you care about the pH of your system and you will be happy if your system resists much change in pH as various factors that can affect pH, change within your ecosystem. So, practically, how should this information affect your actions as you keep your eye on pH? The answer is to conservatively add buffering agents to your system, should you need to affect your pH. A buffering agent is a component of a buffering solution. It moves the pH one way or the other (depending on what agent you use) AND helps to keep the pH from moving. An example of a buffering agent would be any kind of carbonate such as seashells, coral, eggshells, limestone or lime. Carbonates act as buffers by precipitating out of solution at a certain pH threshold. This means that below this threshold, as pH drops, more carbonate dissolves back into solution, while above this threshold, carbonates will precipitate out of solution causing the pH to stay within a certain range — unless the carbonates are exhausted (exhausting the carbonates by adding lots of acid is an example of breaking a buffer). This range depends on the buffer though. This is one of the reasons that aquaponics folks using rock media with lots of limestone have problems with high pH. The pH stays buffered relatively high until all of the carbonates are exhausted and if you're using limestone pea gravel this won't happen for years.

— Nate Storey, Bright Agrotech, LLC

for example, put the agent in a mesh bag, such as a camping dunk bag or nylon stocking, and add this to your fish tank rather than adding the agent directly to your grow bed media or fish tank.

Why might the pH in your system move up? Not having completely inert media, beds and/or tanks is one possible reason. Another is that the water you are adding to your system is not only basic (most tap water is) but is also very hard or alkaline. Alkalinity affects the buffering capacity of your water, and if high can overwhelm the biological forces in your aquaponics systems that are trying to pull the pH down. A third reason is that there may be a spot in your grow bed that has become anaerobic, sometimes referred to as a "dead zone." This is caused by excessive root growth and solids buildup that blocks water flow and oxygenation to that area. If this happens, the area will smell badly and the plants growing there will not look healthy. If you have a sudden, unexpected jump in pH, look for a dead zone and, if you find one, correct it by pulling out the plants in that area, clearing away as much of the sludge as possible and making sure that water is once again able to flow freely through the media.

If you find that your pH has risen above 7.6, the best methods for lowering it can be found in hydroponic stores. My favorite is a product by General Hydroponics called "pH Down." It has been designed to work effectively in a hydroponic plant environment and is free of sodium. Alternatively, you could use other acids commonly found in hydroponic stores like nitric or phosphoric acid. These have an additional benefit — the plants can use the nitrate or phosphate that will be released. Use other acids, such as vinegar (weak), hydrochloric (strong) or sulphuric (strong) acid, but only as a last resort, because adding them directly to your system can be either ineffective or too strong and thus stressful for your fish.

There are a couple of other considerations when adding products to your aquaponics system to control pH. Be cautious when adding products that contain sodium, such as aquarium pH Down. Because the water in an aquaponics system is ideally never flushed out and replaced, the toxic effects of sodium will build up over time and could eventually harm your plants. Sodium levels in aquaponics should never exceed 50 mg/l. Also do not use citric acid as this is antibacterial and will kill the bacteria in your biofilter.

Don't forget that if you are in an emergency situation and your fish are in danger, the fastest way to change the pH of your system is to do a one-third water change of your tank water. But remember; try not to move the pH level more than 0.2 in a day or you will just make a bad situation worse.

On a final note, it is entirely possible that you won't need to adjust your pH at all. Because I have used an inert media, I have an excellent water source and my systems are balanced, I haven't needed to adjust the pH in my systems for over a year and a half. I've heard a theory floated that the red worms might be helping to buffer the pH, but I have no firm evidence of that at this time.

Water — conclusion

In this chapter, we focused on creating an excellent environment for your fish. The keys are to start with relatively pure water that is free of chlorine and chloramines, then adjust it to the right temperature and make sure that it is highly oxygenated. Finally, pay attention to the pH and adjust if necessary. If you follow this recipe, you will create a comfortable home for your fish, as well as optimal water for your bacteria, worms and plants.

Aquaponic Water Rules of Thumb

Purity

- Be sure to "off-gas" chlorine from your water before adding it to your system, or use a chlorine filter.

Temperature

- If possible, select fish that will thrive at the water temperature your system will naturally gravitate toward.
- It is easier to heat water than it is to cool it.
- Attract heat by using a black tank or making it black.
- Retain heat through insulating techniques.

Oxygen

- Dissolved oxygen levels for fish must be above 3 ppm, and preferably above 6 ppm.
- You cannot have too much oxygen in an aquaponics system.

pH

- Target a pH of 6.8–7.0 in your aquaponic system. This is a compromise between the optimal ranges of the fish, the plants and the bacteria.

- Test pH at least weekly, and as frequently as 3–4 times per week, using your API Freshwater Master Test Kit.
- During cycling, pH will tend to rise.
- After cycling, your system's pH will probably drop on a regular basis and need to be adjusted up. If you need to lower pH, it is generally because of the water source (such as hard groundwater) or because you have a base buffering agent in your system (eggshells, oyster shell, shell grit, incorrect media).

Best methods for raising pH if it drops below 6.6:

- Calcium hydroxide — "hydrated lime" or "builder's lime."
- Potassium carbonate (or bicarbonate) or potassium hydroxide ("pearlash" or "potash").
- If possible, alternate between these two each time your system needs the pH raised. These also add calcium and potassium, which your plants will appreciate.
- While they work, be cautious about using natural calcium carbonate products (eggshells, snail shells, seashells). They don't do any harm, but they take a long time to dissolve and affect the pH. So you add it, check pH two hours later and nothing has changed, so you add more. Then the pH suddenly spikes because you have added so much.

Best methods for lowering pH, in order of preference, if it goes above 7.6:

- pH Down for Hydroponics (be careful of using the aquarium version as this has sodium that is unhealthy for plants).
- Other hydroponic acids like nitric or phosphoric as the plants can use the nitrate or phosphate produced.
- Other acids such as vinegar (weak), hydrochloric (strong) and sulphuric (strong) — last resort, as directly adding these acids to your system could be stressful for your fish.
- Avoid adding anything to your system containing sodium as it will build up over time and is harmful to plants.
- Do not use citric acid as this is anti-bacterial and will kill the bacteria in your biofilter.

Above left:
AquaBundance
Aquaponics
system.

Above right:
Outdoor
IBC Tote
Aquaponics.

Left:
Water Feature
Aquaponics.

Right: Backyard Aquaponics system.

Below left: Filling the grow beds with gravel.

Below right: Bathtub Aquaponics

JOHN BURGESS – MEMBER "RUPERTOFOZ"

IMAGE COURTESY OF ECOFILMS AUSTRALIA

NICK LA HAISE AT SYZYGY HOUSE

Left:
Balcony
Aquaponics.

Below left:
Aquaponic
lettuce.

Below right:
Tilapia.

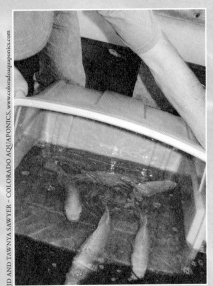

Above left:
Closet system.

Above right:
Catfish.

Right:
Greenhouse
Aquaponics.

Above left:
Aquaponic
corn.

Above right:
Practical
Aquaponics.

Left:
Formal
Aquaponics.

IMAGE COURTESY OF THE AQUAPONIC SOURCE, INC.

JD AND TAWNYA SAWYER – COLORADO AQUAPONICS. Www.coloradoaquaponics.com

Above left:
Luna fishing.

Above right:
Red Worms in
Hydroton and
Lava Rock.

Right:
Greenhouse
Aquaponics.

IMAGE COURTESY OF THE AQUAPONIC SOURCE, INC.

*Left:
Basement
Aquaponics.*

*Below left:
Duckweed.*

*Below right:
Aquaponic
chili peppers.*

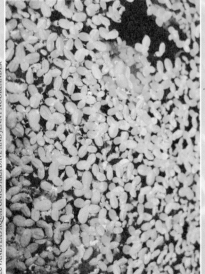

RIK KRETZINGER'S AQUATUBE SYSTEM IN NORTHERN CALIFORNIA

TCLYNX, AQUAPONIC LYNX LLC www.aquaponiclyns.com

Above left:
AquaTubes.

Above right:
Banana tree
in Aquaponics.

Right:
Mixed
Aquaponics.

KOBUS JOOSTE, AQUAPONIC SYSTEM DEVELOPMENT

Section 4

The software

11

Fish

"I see aquaponics as a technology that offers sources of fresh fish and organically produced vegetables, giving every consumer an opportunity to become a producer. As a chef, this offers a whole new level to seasonal variety, produced without hidden transportation or fertilizer costs. The only additive to the system is the fish food, which pays back with both vegetable and seafood. With a home system, there is a deep level of satisfaction with seeing the ecosystem in excellent balance, with a minimum of work, and having something way cooler than a TV in the living room. Hydroponics is clever, aquaponics is smart."

— Dan Brown, Denver, Colorado

When I am asked why someone would grow aquaponically, I go through the usual reasons that we probably all use, or will use, to justify the time and money we spend on our obsession. We cherish organic produce that is sustainably grown, we want to be able to garden anywhere regardless of soil quality, we do not want the normal mess or weeds, etc. I often find myself wrapping it up with the heartfelt exclamation, "But the real reason is that it's our landscaping and think about how I enjoy the flow really static and, worse, it doesn't greet me in the m

Fish are clearly aquaponics system. They don't ask much of us: ox and temperature range and food. In return, they give our plants life. And perhaps the best part is how excited they get when I go into my greenhouse and it's feeding time. The tilapia all crowd around the spot where I throw in the food, waiting at the surface. Then the frenzy ensues as I feed them, providing a little taste of the wild kingdom in my otherwise tranquil greenhouse.

In this chapter we will discuss what you should consider when selecting your fish, where to find the fish you select and how to raise them successfully.

Don't worry if you have never before cared for fish. In many ways, aquaponically grown fish in large tanks are much simpler to care for than pet-shop fish in a glass aquarium. Filtration is automatically provided as a benefit of the system. Water temperature and pH become more stable the larger the body of water. Disease is much less prevalent in aquaponically grown fish (researchers are beginning to think that plant roots actually give off a natural antibiotic). And there is no need to keep the glass free of algae and the gravel and tank toys looking pristine. In fact, we want bacteria to build up on the sides of the tank, so hopefully you aren't burdened with the shackles of a system on display.

How many fish can I grow?

This is one of the first questions most aspiring aquaponists ask. If you are considering growing fish for food, how many you can grow goes right to the heart of a cost/benefit analysis. Just how much can you reduce your family's grocery bill through aquaponics?

The generally accepted rule of thumb for a home, media-based aquaponics system is one pound (500 g) of fish for every five to ten gallons (20 to 40 liters) of fish tank water. By comparison, a professionally monitored, mechanically filtered recirculating aquaculture facility will stock as densely as one pound of fish for every one and a half gallons (six liters) of water. So what drives the difference?

The commercial facilities are constantly monitored by professionals using computerized equipment, with intense ox   huge water and energy consumption. The solid wast   d from the system. They also "grade" the fish b   r than average) into tanks with other similar-   to ensure that each tank contains the same s   In home aquaponic gardening we actually   so we can harvest fish frequently over a long

Since it isn't particularly convenient for home aquaponic gardeners to think about stocking densities in terms of pounds of fish, an ancillary rule of thumb has been developed over time. It says that you can think of stocking your tank in terms of the number of fish. Conveniently, the rule is about one fish for every five gallons (20 liters) of water. This works because when you first start your garden, the biofilter that evolves from the development of

Thoughts on stocking density

Fish will release more waste as they get bigger. You should stock your system with the future average maximum fish density you require to support the growth in your grow beds. When the fish reach maximum density, it doesn't matter if you still leave them in for a while. In fact, what you should do is aim to get to your average daily feed in/out as fast as possible; the fish don't matter that much. So I recommend people purchase fish numbers that will closely match the final harvest weight and the maximum density.

Example: if you have a 1,000-litre tank and want a density of 20 kg/1,000 l and you harvest at 500 g, then you need 40 fish. I would purchase 50 at the stage where I put them in the tank. This system would need about 200 g of fish feed per day, so it will take time to get the feed up to that level. Beginners are better off doing this and having some patience rather than going for high fingerling numbers and then trying to get rid of them later. 50 x 5 fish only = 250 g of fish. If they ate 1 percent of biomass per day, that would mean you only add 2.5 g of feed!! However, fish this size will eat as high as 7 percent of their weight, so you could add 10 g per day; still too low! BUT, most systems are skewed to excess nutrients, so you can grow some plants on this feed level.

Most people are tempted to stock more than this rule of thumb suggests. I strongly urge you to resist this temptation. Aquaponics forums are full of stories of new gardeners who have tried to push these limits and have unfortunately suffered disasters. Your grow bed biofilter can only process a finite amount of fish waste at any point in its development. Think of this process like creating a fine wine. The fish waste is represented by the grapes that are pulverized and then put through a yeast fermentation process (the biofilter) before it becomes wine. With inadequate fermentation you just have soggy, wet grape mash. The magic happens when the grapes are in balance with the yeast. The same works in aquaponics. Your plants will only benefit from food (the wine) that the bacteria biofilter produces. If the biofilter becomes overwhelmed, you will jeopardize your fish from lack of filtration, and ultimately your entire system. Please stick to the recommended ratios.

— Dr. Wilson Lennard,
Aquaponics Solutions, Australia

the nitrifying bacteria in your grow beds is immature and can only handle a fraction of the waste it will be able to deal with when it is fully mature. For example, if you start a newly cycled 100-gallon (400-liter) system with 20 fish fingerlings, the fish and bacteria biofilter will grow up together. With just fingerlings, you will not have much ammonia to convert to nitrates. But in nine months to a year's time, when your fish are likely reaching maturity, your biofilter will also be mature and able to effectively convert the higher ammonia and solid waste load from fish that are at or even over a pound (500 g) in size.

What type of fish can I grow?

This question should actually be asked early on in the design of your aquaponics system. The answer could impact both the size and placement of your system. Where you place your fish tank may be especially important because different fish require different temperature ranges to thrive and tank placement may dramatically affect water temperature. More on this below.

The one hard and fast rule with regard to fish selection in aquaponics is that you must use freshwater fish. This is because most plants won't tolerate extended exposure to the sodium in salt water.

Here is a quick list of some popular fish grown in home aquaponics in North America:

- Tilapia — when people think of aquaponics they typically think of tilapia. This is by far the most frequently used fish in aquaponics. Why? Because it

Tilapai. Fishgen

is easy to grow, likes warm water, does not have high oxygen requirements, reaches harvest size (approximately 12 inches / 30 cm and one-and-a-half pounds / 700 g) in 9–12 months and tastes delicious!

- Goldfish — probably the second most popular fish because again, it is easy to grow and is widely available. This is generally the fish of choice for those who aren't interested in eating their fish.
- Catfish — a popular choice for those who live in states where tilapia is illegal (e.g., Florida). Again, an easy fish to raise and it grows quickly.
- Koi — for those who want to take their goldfish hobby to the next level with a hardy, beautiful fish that can ultimately reap awards and very high prices.

In case the short list above is not enough for you, others raise shrimp, barramundi, pacu, perch, trout, oscars …and yes, even freshwater lobster. Just remember to take into account the considerations covered below before making your final selection.

The rest of this chapter is devoted to helping you select the right fish for your goals and growing environment as well as covering how to acquire and care for the fish you select.

The next question you need to ask yourself is "what do I want from my fish (other than their waste)?"

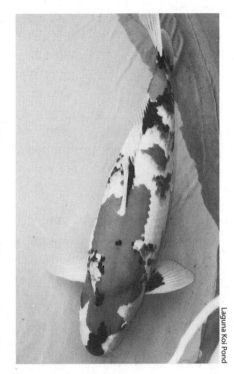

Koi.

Are you looking for food, fancy or fun?

Most aquaponic gardeners want their fish to become food for their table. Aquaponics is especially appealing because it simultaneously, in one system, produces both plant and protein crops. The notion of pulling a fresh tilapia, perch or catfish straight from the water just feet from a hot grill is enchanting. No waders, bug spray or camping trip required! The result — you get the freshest possible flavors while knowing exactly how your fish spent its life, what it ate and how it was harvested. Plus, you can also feel good that you haven't contributed to the serious problem of overfishing our oceans or polluting them through offshore aquaculture operations.

Some aquaponic gardeners, however, are not interested in eating their fish but rather in raising beautiful, fancy fish such as koi. Some want unusual, fun fish; perhaps pacu, the vegetarians of the piranha family that grow two to three feet (60–90 cm) long, eat kitchen scraps and like to be stroked by their humans.

If you honestly don't have a strong opinion about the type of fish you grow, I recommend goldfish. They are hardy in a wide range of temperatures and conditions, will do a terrific job producing waste for your plants and grow slowly, so they live comfortably in tanks ranging from small aquariums to large tanks.

To eat or not to eat? That is the question

When you select the fish for your aquaponics system, the first decision you need to make is whether they are destined for your dinner plate (for example tilapia, catfish or trout) or are strictly ornamental (for example koi or goldfish). Many people are drawn to aquaponics because of the allure of growing their own edible fish.

I've found, however, that as I've lived with my aquaponic systems, those lines are no longer so clear.

When I started in aquaponics, from day one I had visions of a dinner party with freshly caught tilapia as the centerpiece. Problem is, my fish and I go through a lot together and I'm afraid we have bonded. I admire their ability to endure the experiments I conduct to learn from them. I enjoy their elegant beauty as they float through the tank and their hunger-induced enthusiasm for my greenhouse visits. While I still joke about their future, I joke with less conviction these days. To eat or not to eat?

Consider the pacu. It has a reputation for being a wonderful food fish — internet stories of barbequed pacu ribs made my mouth water. They are being raised on farms in South America for this very purpose. Pacu are vegetarian members of the piranha family that eat just about anything (table scraps) and are incredibly fast growers that will reach 18 inches (45 cm) in about a year. Great metabolism equals lots of plant food and a fast journey to dining nirvana. And yet …. they apparently also love to be stroked and will greet you at the tank's edge like a koi. To eat or not to eat?

I discussed all this with the manager of a local pet store who has a long history of raising fish. He pointed out that many fish they sell as ornamentals are actually treated as food fish in South America. Tilapias are cichlids, and he had at least five varieties of "fancy" cichlids on display in fish tanks for fancy prices at his store. Oscars are the in the same category — pets in the US are often dinner plate fodder in South America.

So when you are figuring how to stock your aquaponic system and are trying to decide "to eat or not to eat," my advice is not to worry about it too much. You can always change your mind.

Next, consider temperature ranges, oxygen levels and food requirements to hone your thinking on which fish to raise. Read on.

Temperature

Fish are cold-blooded animals, meaning they take on the temperature of the water in which they live. One of the reasons fish are so efficient at creating body mass is that they don't use energy to keep their body temperatures stable. Mammals use approximately 80 percent of their caloric intake to keep their bodies at optimal temperature; fish use none. Consequently, fish have a feed conversion ratio (the mass of the food eaten divided by the body mass gain) of 1.5, compared to 7 for warm blooded cattle. (Another reason for the 1.5 ratio is that fish are supported by the water in which they swim while sheep and cattle expend energy just to stand up.) This is also part of the reason why fish and other cold-blooded animals don't take on diseases such as *E. coli* and *Salmonella;* their bodies don't provide the nice warm environments that viruses, bacteria and parasites need to live. However, if you do not match your fish choices to the water temperatures they will experience, you will not achieve optimum growth and you may end up with unhealthy fish.

Different fish have different water temperature requirements depending on their natural climates. For example, tilapia originated in the lake waters of Africa and therefore evolved to thrive in warm water (above 60°F/16°C). Trout, on the other hand, originated in the streams of North America, Northern Asia and Europe and, as such, require cool to cold water (55°F/13°C and below). Goldfish can tolerate a wide range of temperatures, and readily survive ice-covered outdoor ponds in cold winter climates. So the challenge is to select fish that will thrive in the water temperature you will be able to provide.

The water in your fish tanks will naturally stay in a set temperature range presuming you neither directly heat nor cool the water yourself. The range will depend entirely on the environment where you locate the tanks. Will they be in a greenhouse that is heated in the winter and gets extremely hot in the summer, or perhaps in a basement that stays cool all year round, or on your back deck in the summer and indoors in the winter, or perhaps outdoors year-round in a temperate climate where they may face only a couple nights of frost but will regularly get down into the 40s (below 10 °C) for a few months?

With ideas for which fish you want to raise and a sense of the water temperatures you will most easily be able to achieve, you can match your fish choices to your water temperatures and see what works best. You may decide that you will want to modify the water temperature in one or more of your tanks for some or all of a yearly cycle. If you do, you should consider that it is far less expensive to heat water than to cool it, both in equipment and in ongoing energy costs, and the fewer degrees you need to move the water temperature in one direction or the other, the less power you will need. Try to match the temperature range in which your fish thrive to the temperature range that your tank water will naturally acquire as well as you can.

I have heard that some aquaponists attempt to grow a warm-weather species (e.g., tilapia) through the summer months with the notion of switching to a cool-weather species (e.g., trout) for the winter. While this sounds ideal in theory it would be tough to successfully execute since both fish take 9–12 months to fully mature. If, however, you have reliable access to more mature fish that are already halfway to a harvestable size then by all means give this strategy a try.

The chart on page 142 lists the temperatures at which several fish varieties thrive. As discussed above, this is an important consideration because when they are in their most comfortable temperature zones, fish reach their highest metabolic rates, eat the most food and subsequently provide the most fertilizer for your plants. It is the same with us. If we are very cold or very hot, the last thing we want to do is eat. In addition, keeping your fish at their ideal temperature helps build their immune systems so they can ward off any possible disease.

And a final word of caution about fish and water temperature from Dr. Lennard: "The biggest issue with temperature and fish are fast changes in temperature. Most fish can handle wide temperature ranges and live, but if the temperature changes more than 2 °C (3–4 °F) in a 24-hour period, they can be upset. If you experience wide temperature swings over short time periods, you should immediately stop feeding. Most fish die from gut bacterial problems associated with temperature swings."

Eating habits

Next consider what your fish will need to eat. Are the fish you are considering omnivores or carnivores?

Carnivores (trout, bass, perch and oscars) require a high-protein diet that is difficult to achieve without purchasing a high-quality commercial feed specifically formulated for carnivorous fish. For example, Purina has a separate product line from its AquaMax fish feed called AquaMax Carnivore. If you like the simplicity of just tossing in a handful of packaged feed a couple times a day, this won't be a negative for you. If, however, you would take pleasure in raising some of your own fish feed, you will be limited to grubs and worms with carnivorous fish.

Another consideration with growing carnivores is that they are, well, carnivores. While almost all fish will show some proclivity toward nibbling on their neighbors, especially the young and the weak, other fish are the food of preference for carnivorous fish. This means that not only can you not mix them with other species in your tank but you should also be sure that all your carnivorous fish are approximately the same size or they will snack on each other.

On the other hand, omnivorous fish (tilapia, catfish, pacu, koi and goldfish) generally coexist well with members of their own species and with other omnivorous fish species. Because they eat a varied diet, you can reduce your feed costs substantially by growing much of your own feed. Plus, if you are raising fish for you and your family to eat, using your own feed reduces the concern about possible contaminates coming from outside feed sources and gets you even closer to creating a self-sustaining source of protein.

Oxygen needs

All fish require oxygenated water, but some require more than others. Again, consider the fish's native environments. Tilapia and catfish hail from ponds, lakes and marshlands and so are used to relatively low amounts of oxygen and low water quality in general. Trout, on the other hand, evolved in streams and brooks sourced from clear mountain snow runoff. They are therefore happiest in highly oxygenated waters.

In fact, all fish species do best in a highly oxygenated environment. Instead of focusing on the exact oxygen requirements for the fish you are considering, think in terms of safety margins. If you have stocked with a low-oxygen fish species and something goes wrong with the aeration system in your fish tank

	Tilapia	Trout	Catfish	Bass	Goldfish	Koi	Pacu
Edible?	yes	yes	yes	yes	no	no	maybe
Temperature range (°F)	60 - 95	35 - 68	35 - 95	40 - 90	35-90	35 - 90	60 - 95
Optimal Temperature (°F)	74 - 80	55 - 65	75 - 85	74 - 80	65 - 75	65 - 75	74 - 80
Carnivore or Omnivore?	O	C	O	C	O	O	O
Mature Size	1.5 lb	0.8 lb	1.25 lb	1 - 3 lb	4"	20 lb	60 lb
How long to reach maturity?	9 - 12 mos	12 mos	12 - 18 mos	15 - 18 mos	3 yrs	3 yrs	4 yrs
Oxygen Needs	Low	High	Low	Low	Low	Low	Low

Fish comparison chart.

for some period of time, you will have much more time to recover and be less likely to face disaster than if you chose fish requiring high levels of oxygen. We discussed ways to affect water oxygenation in Chapter 10: Water.

The above chart highlights the species-specific parameters for the criteria we have discussed.

Sources of fish

Now that you have thoughtfully analyzed what fish you want fueling your aquaponics system, the next question is "where do I find them?"

If you have selected ornamental fish your quest will be a fairly simple one. Most pet stores carry a variety of goldfish, and even oscars and pacus can be found with a little bit of searching. If you have chosen koi, you can find fingerlings at a specialty aquarium shop or mail order them through a website, depending on how particular you are about the exact variety of koi. The source you choose will depend in part on whether these are just fish for fun, beauty and aquaponics production or if they are to be part of a serious showing and breeding enterprise. There are a huge number of koi varieties, and each can be graded into several classes according to size, pattern, color and social behavior. Within these parameters koi can cost anywhere from a dollar to thousands of dollars each.

Sourcing edible fish

If you have chosen an edible fish, sourcing is a little more challenging. First, I recommend that you start with your local Fish and Game Department to find out which fish are allowed or, perhaps easier, which fish are banned in your state or province. Many states have outlawed raising tilapia, for example, because they have been released into the waterways, are successfully breeding and are now considered a threat to native species.

Once you have established which fish you are allowed to keep, ask the same Fish and Game employee, or search the internet, to locate a fish hatchery within your state that is willing to supply the type of fish you want at the quantity that makes sense for your fish tank size. This is a good option for several reasons. First, there will be no doubt that hatchery-raised fish are legal in your state. Second, the fish you find might be a subspecies that is particularly acclimated to your region. Catfish are a great example of this. Catfish purchased in Florida will not survive cold winter temperatures, but catfish purchased from a supplier in Ohio will. Finally, it is just a good idea to get to know and support your local hatchery as it could be a terrific resource for your fish-specific questions.

If you can't find a hatchery, you will need to resort to finding a vendor on the internet. This is an expensive option because you will be paying for specialized packaging and overnight shipping of a heavy box largely filled with water. Also, you, and not the merchant, will be responsible for being sure that the fish type is legal in your state or province. On the plus side, if you get a breeding mix of male and female fish, you may only need to make a single purchase because from that point on, you will have your own supply of baby fish.

All-male tilapia

This past summer a photographer came to our home to take product shots of our new AquaBundance aquaponics system. Since it was a bit of a drive for him, he and his wife decided to make a day of it and brought their young son. He was of course instantly drawn to the fish, and as he peered into the tank he exclaimed, "Look at all the babies!" "Impossible," I thought; "I have all-male tilapia stock." But I humored him by looking into the tank myself. I'll be darned. Floating on the surface of the tank was a miniature school of about thirty tiny fish. Those boys had had babies!

As background, most commercially raised tilapia are male because males grow bigger faster, and reproduction is just plain messy, no matter what the species. Babies lead to fighting, territorialism and the tank quickly gets overpopulated.

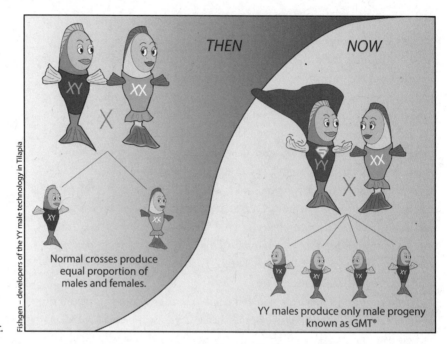

THEN

NOW

Normal crosses produce equal proportion of males and females.

YY males produce only male progeny known as GMT®

Fishgen – developers of the YY male technology in Tilapia

All male tilapia graphic.

There are two main methods used in the US today to ensure all-male tilapia stock. The first, most frequently used method is called "direct hormonal sex reversal." Basically the young fry are flooded with male hormones at an early age causing them all to grow up male. The second method, and the one I find more palatable, is called "genetically male tilapia" (GMT). Basically a super male is produced with YY chromosomes (i.e., two male chromosomes) so when he reproduces he can only produce another super male or a male. The wonderful graphic above from FishGen.com will help explain what is going on.

Last December I got 100 "all male" tilapia that had been sexed using GMT, but since the inventors claim that the process is "95 percent effective" I guess I shouldn't have been surprised that a few girls slipped in. I enjoyed the babies and the fact that Mother Nature always seems to find her way around our efforts to thwart her.

Breeding tilapia

Perhaps you have decided to seek out both male and female fry so you can have a perpetual supply of edible fish. When it comes to breeding tilapia,

Tilapia breeding

Tilapia will reproduce to the point of danger and even overwhelm the biofilter if your adults are well fed, and the young can find any refuge in the tank. If hungry, the adults will often cannibalize to some degree, but rarely will they control their own population. They prefer a lot of other food items over their own young if they are readily available. Non-spawning adults do have a seemingly unsatisfiable appetite for eggs though. Just one female will typically produce about 200–1000 eggs per spawn, and she'll spawn every 4–5 weeks or so if conditions are decent enough in the tank ("decent" is pretty easy for tilapia). Even with low survival, that's still a lot of tilapia recruitment. So... in the average home system, even a single female could have a tank filled to the brim with young tilapia in no time at all... perhaps just a single spawn! Interestingly, the juveniles from previous spawns will actually be the most cannibalistic fish in the tank. Young tilapia have a huge appetite for high-quality protein, and tilapia fry are a great source of protein in their eyes. They'll eat any sibling they can fit in their mouth ... something worth keeping in mind if you are actually trying to grow out some of them. Grading by size is literally a full-time job at our hatchery. You have many options when it comes to preventing excessive spawning and/or recruitment. Here are a few:

- Increase adult fish density — Crowded tilapia rarely pull off successful spawns. Males prefer to have some room for their nest area. They will chase off anyone they consider an intruder. In a heavily stocked tank, they are too busy chasing non-spawning fish away from their nest site to have much time to actually pull off a spawn. Further, even if they do manage to court a female and get her to release eggs in the nest site, the huge population of non-spawners will attempt to swoop in and eat the eggs... and they'll do so with great success. Of course, you'll need to increase biofiltration capacity to suit any increase in fish density, and you still need to keep in mind that all tanks have a maximum carrying capacity at some point. High densities can be risky. A lot more risk for water quality parameters to turn south quickly.

- Decrease their "tank" size without decreasing total water volume — One way to do this is to use fish cages in your tank with mesh bottoms. Another way is to reduce the size of your fish tank while maintaining an auxiliary tank or sump equal to the water volume you eliminated from the fish tank. No spawning area available means no spawning will take place.

- Introduce a few predator fish. The right selection will not bother your adult tilapia, but will keep the "herd" trimmed for you by eating most of the young. There are all sorts of fun predator fish you could keep with them. I have a research report that concluded largemouth bass and tilapia kept in a polyculture RAS improved growth rates substantially, compared to the monoculture control tanks of ☞

bass only or tilapia only. Similar studies have produced similar results using catfish and tilapia in polyculture systems.

- Separate the males and females. This obviously works, however, it can be pretty time-consuming in large systems, and only works as well as the person's ability to correctly ID the sexes. In most home systems, it's not a huge deal to handle 50–200 fish though.

- Stock single sex fish — Stocking "all male" fish, for instance. Just keep in mind that hormones are used to produce "all males" unless they come from a genetically male cross. Since most aquaponics folks are growing some of their own food to avoid hormone-treated foods, this doesn't usually jive with our goals.

- Reduce light. Tilapia spawn less frequently and less successfully when provided with more dark hours than light hours.

- Reduce temperatures slightly. Tilapia tend to spawn at optimal levels at temperatures between 78 and 84 °F (26–29 °C). At 72 °F (22 °C), most strains virtually stop spawning altogether, though they also grow slower... so not really the best solution in most cases.

- Personally, I like the idea of using predator fish. It's not fail proof by any means, but it works well enough to control the population, and it can be a lot of fun too. You are able to maintain optimal conditions for fast growth, while controlling the population. For our own home systems, I've used largemouth bass, hybrid striped bass, crappie, yellow perch, catfish (they are far more piscivorous than most people realize), oscars and Jack Dempseys... with varying results. The most successful were the largemouth bass, followed by oscars and Jack Dempseys. The LMB [largemouth bass] results jive with several recent studies, but the oscars and Jack Dempseys were purely from my own fairly non-scientific trials conducted just "to see what would happen." They worked great for me, but your results could be different. The best predator fish to select are ones that have been raised entirely on live food and never tasted pellets (you want them eating fry and small fingerlings, not pellets meant for your tilapia). That's not always an easy fish to find.

— Kellen Wessenbach,
White Brook Tilapia Farm

the challenge isn't how to get them to breed, but rather how to handle all the fish that are hatched. I turned to my trusty fish guru to give advice on how to handle this embarrassment of riches see ("Tilapia Breeding" on the previous page.)

Introducing fish into your aquaponics system

The day your fish arrive is an exciting one. This is the moment when you truly feel like you are moving from hydroponics into aquaponics. There are some steps you should take, however, to be sure that the transition is a smooth one and that your fish get a good start in their aquaponic life.

Be sure your system is ready — I tell you how to start up or "cycle" it in a later chapter (Chapter 14: Cycling). For now please note that that process needs to be completed and your system should be mostly free of ammonia and nitrites.

Match pH — Be sure to ask your fish vendor the pH of the water the fish have been raised in, or check it in the bag in which they arrive. Fish are very sensitive to quick pH swings and will be stressed if the pH of the water in their bag is more than two-tenths of a degree different from the pH in their new tank. Be sure to adjust your pH to match that of the fish before you add them to the water, even if this is not a pH that is optimal for aquaponics (6.8–7.0). You can slowly adjust your pH once the fish are in their new home.

Match temperature — Because fish are cold-blooded creatures, their body temperatures fluctuate with the temperature of their environment. They will be stressed if they are suddenly introduced to water that is more than a couple degrees colder or warmer than the water they are currently in. The best way to acclimate the fish to the temperature of the tank is to let the bag they are in float in the tank water until it reaches the same temperature. This usually takes ten to fifteen minutes, and can be accelerated by occasionally adding some water from the tank into the bag.

Feeding your fish

When and how often?

Fish tolerate a wide range of feeding schedules very well. They actually adjust their metabolism to match the availability of food. If you want your fish to grow quickly, or you have fewer than the recommended number and need to produce more food for your plants, go ahead and feed them often. Commercial aquaculture operations feed their fish as often as once an hour so as to feed their adult fish as much as 1 percent of their body weight in feed per day, and closer to 7 percent for juvenile fish. If, however, you are in an overstocked situation, or had an insect outbreak in one of your planting beds so you need to replant but only have small seedlings available, or for any number of other reasons, you need to "dial down" the amount of fertilizer your fish are producing, simply feed them less. I often only feed my fish once a day, in the morning when I go to check on everything. Sometimes, however, I'll feed them again in the evening if I need to go to the greenhouse to pick some produce for dinner.

The best rule of thumb is to only feed your fish as much as they will eat within five minutes. After five minutes, remove the remaining food from the tank with a fish net. Based on your fish's behavior at that moment, you will soon be able to judge just how much food to toss in, and will no longer need to wait five minutes to see how much they eat.

The time you spend observing your fish when you feed them is very valuable. For this reason you should avoid automatic feeders unless you are going on a short vacation and have no one to take care of your fish. Automatic feeders aren't inherently bad, but if your fish stop eating, something has likely gone wrong with your system and you will not receive this critical signal if you aren't there at feeding time. Fish may stop eating for a wide variety of reasons including water temperature being outside of "thriving" range, pH being outside tolerable range, too much ammonia and/or nitrites (a fish may have died in the tank, giving off excess ammonia), too little oxygen, stress or a disease. All of these issues are easily corrected if caught early, and potentially fatal if not.

Commercial feed

Commercially prepared fish feed is an excellent source of nutrition for your fish. There are typically two types of fish feed sold in the US — omnivorous and carnivorous — and they vary mostly in their protein content. Within these types, you can select feed according to the stage of growth your fish are in. Again, the difference will largely be the amount of protein in the feed, although you will also notice a difference in pellet size. Not surprisingly, as the fish get older, the size of the feed pellets gets larger.

Fish feed is comprised of proteins, fats, minerals, carbohydrates and other nutrients that a fish in the wild would have in its normal diet but are lacking in captivity in what is essentially a wet desert. The source of these nutrients is usually fish meal, corn, soy and other animal byproducts.

The protein in fish feed comes mainly from fish meal. Fish meal can come from fishery wastes associated with the processing of fish for human consumption or from specific fish (herring, menhaden and pollack) that are harvested solely for the purpose of producing fish meal.

There is currently intense debate within the aquaponics and aquaculture communities about the wisdom of adding to the serious problem of overfishing our oceans by feeding our farm-raised fish more fish from the ocean. Thankfully, many experts are conducting groundbreaking research to develop protein substitutes for fish meal. An article in the *Aquaponics Journal* (Nelson, 2010) highlighted three companies creating protein sources. At Ohio State University, aquaculturalists are exploring the feasibility of using soybeans to replace fish meal and soon plan to test the product on yellow perch. Scientists at the Agricultural Research Services and Montana Microbial Products have teamed up to produce a barley protein concentrate that can be fed to trout, salmon and other commercially produced fish. Finally, in Idaho Springs, Colorado, Oberon FMR has signed a deal with Miller–Coors to use 5,000 tons of beer sludge in combination with other ingredients to produce 6,000 tons of fish food flakes.

All fish feed, especially brands that use more natural ingredients and fewer preservatives, have a limited shelf life and are best stored in a cool, dry location.

Homegrown feed

Supplementing, or even entirely substituting your own, homegrown fish food, can be personally satisfying, save money and further decrease the environmental footprint of your aquaponics system by further closing the input loop. The following is a list of some of the feed that can be easily cultivated for most omnivorous fish:

Duckweed tank.

- Duckweed — This fast-growing aquatic plant doubles in mass every day when in its ideal environment. It is also more than 40 percent protein (more than soybeans) and efficiently removes contaminants from, and adds oxygen to, the water. Duckweed grows best in dappled sunlight in relatively stagnant water with some fish waste. The key is to keep it in a separate tank from your fish or they will just eat it all!

- Worms — Earthworms, sludge worms, bloodworms and composting red worms (aka red wrigglers) all make excellent fish food. The challenge is to grow enough of them to be more than an occasional, although probably very appreciated, treat for your fish.
- Black soldier fly larvae — The black soldier fly (BSF) is a native of the North American continent and can be found in many parts of the United States and throughout the world. It is exceptionally active in the southeastern United States from April to November. Their grubs are considered beneficial scavengers in nature, and help to digest and recycle decomposing organic material including carrion, manure, fruits and decaying plant waste. Their association to humanity is limited to compost piles, facilities producing manure and poorly serviced toilets. Unlike the common house and fruit flies, black soldier flies are not commonly found in association with people (picnics, kitchens, residential buildings, etc.). While the mature fly has a short lifespan of only five to eight days, the female can lay over nine hundred eggs. Those hatch in about a hundred hours and, if conditions are right, will mature in two to four weeks. During this stage, the larvae make excellent fish food, as well as robust compost consumers. A product called the BioPod is specifically designed for BSF food composting and through clever design, has the added benefit of self-harvesting the larvae.

Karl P Warkomski, CompostMania.com

Black soldier fly.

Karl P Warkomski, CompostMania.com

Black soldier fly larvae.

• Other kitchen and garden scraps — Omnivorous fish like most bland-tasting, non-flowering plants and even some fruit. I've found that my tilapia especially appreciate lettuce that is getting slimy and no longer fit for human consumption. I've been told that they will eat untreated grass clippings (no weed killer or pesticides!). They get very excited about the uneaten tops of strawberries. Experiment! If they don't eat what you have given them in five minutes, and it is their normal feeding time, remove it from the tank and chalk it up to experience. What have you got to lose?

I recommend that you consider any one, or a combination, of the feeds above as a supplement to a commercial fish feed. Nutrition for living beings is a complex subject, especially in captivity. Just as we wouldn't feed our dogs or cats just one or even a couple foodstuffs, consider feeding your aquaponic fish a varied diet that includes a reputable commercial feed. Think of it as a vitamin tablet for your fish to ensure that both they, and ultimately your plants, are getting all the nutrients they need.

Harvesting your fish

Chances are good that part of your attraction to aquaponics is being able to grow your own fish for your dinner plate. With this aspiration comes the need to "dispatch" your fish as humanely as possible when both they, and you, are ready.

First, I recommend that you "purge" your fish by putting them in a tank of clean water and not feeding them for a few days before harvesting. While not absolutely necessary, this process gives the flesh a better, cleaner flavor.

While never pleasant, there are several ways to dispatch a fish. The most obvious is through asphyxiation, by pulling them out of the water. This is, however, not very humane because a fish may endure quite a long period of suffering before dying from lack of oxygen. More immediate methods are either to hit it on the head with a blunt object or insert a knife between its eyes. The best way I've heard of, however, is to plunge the fish into an ice water bath. The fish quickly becomes unconscious from the shock of the temperature change and dies shortly thereafter. While this method won't be effective for cold-water varieties such as trout, for warm-water species it is both a quick death for the fish and less traumatic for you than trying to hit a wriggling fish on the head!

Fish summary

Fish are the fuel, the fun and the fascination of any aquaponics system. They are what your neighbors will come over to see, your kids will beg to feed and your dog will want to play with. Choose your fish thoughtfully, source them from a reputable vendor, treat them well and they will reward you with a lively, healthy growing environment, and perhaps even some excellent meals.

Aquaponic Fish Rules of Thumb
Stocking density

- 1 pound of fish per 5–7 gallons of tank water
- Fish selection should take into account the following:
  - Edible vs. ornamental
  - Water temperature
  - Carnivore vs. omnivore vs. herbivore

Oxygen needs

- When introducing new fish into your system:
  - Be sure your system is fully cycled
  - Match pH
  - Match temperature

Feeding rate

- Feed your fish as much as they will eat in five minutes, one to three times per day. An adult fish will eat approximately one percent of its body weight per day. Fish fry (babies) will eat as much as seven percent. Be careful not to overfeed your fish.
- If your fish aren't eating they are probably stressed, outside of their optimal temperature range or don't have enough oxygen.

12

Plants

"I find aquaponics horticulture extremely interesting and something I could see myself doing for the rest of my days. I used to garden when I was younger but fell out of practice. I forgot the joy it can bring you. I went to college for a few years but didn't really find anything that 'fit' ME. I really think this aquaponics is ME and I could contribute more than you will ever know."

— Teddy Malen, Brookfield, Wisconsin

Fish differentiate aquaponics from other forms of gardening, but the plants supply most of the food. Your goal may be to feed your family fish once or twice a week, but that much fish can provide enough fertilizer to grow enough plants to have a significant impact on your grocery bill.

In this chapter we will discuss what kinds of plants can be grown in aquaponics, what those plants need to survive and thrive, how to start plants for aquaponics, how to transplant them into your system and how to prevent, identify and fix plant problems.

What plants grow best in aquaponics?

Actually, I should call this section "what doesn't grow well in aquaponics?" because we would have a much shorter list. There is only one type of plant that I know of that absolutely does not grow well in aquaponics and that is any plant that requires a pH environment much above or below neutral 7.0. Examples of this are blueberries and azaleas, which prefer acidic soil (below

7.0), and chrysanthemums, calendula and zinnias, which prefer basic soil (above 7.0).

I'm often asked if subterranean plants, or root vegetables, can grow in aquaponics. The answer is yes, and no. They will grow, and often very productively, but you might not recognize their final mature shape as that of a carrot or a potato. This is because a carrot or potato will have a harder time expanding into its mature shape in gravel than it would through soil. But the taste will be just as good, and perhaps you will discover new, amusing twists to your traditional root vegetables.

Beyond those two considerations, you can grow anything in aquaponics that you can grow in soil. I know of aquaponic gardeners who have even grown papaya and banana trees! Certainly salad greens, tomatoes, peppers and strawberries all thrive side-by-side in an aquaponic media bed. I've grown herbs of all kinds, from prolific basil plants to low-water thyme. Flowers of all colors and sizes also do very well in aquaponics. You are only limited by your imagination and your seed library.

Growing plants in aquaponics

Without getting into a detailed botany lesson, I'd like to spend some time on how life for plants growing in an aquaponics system is both the same, and different, from the life their soil-based counterparts lead.

You can probably recite the basics of what plants require in order to grow from recollected school memories or more recent home gardening experiences. See if these sound familiar.

Water

While this may seem self-evident, the method for delivering water to the plants is one of the main differences between aquaponic and soil-based gardening. In nature, plants are dependent on rain and sometimes groundwater sources for the water they need to grow. A plant community develops around the delicate balance between the average rainfall of a microclimate, the ability of the soil to retain moisture, the native plants' need for water and their ability to retain it in times of drought.

Along comes man's discovery of agriculture and soil gardening. Now the gardener can select plants based on the season and his food needs and tastes, and modify the environment to suit the needs of those plants. The soil is amended to better retain water, and a lack of sufficient rainfall is mitigated through irrigation.

While agriculture and gardening have important advantages over purely natural ecosystems in watering reliability and crop selection, there are still problems. First, because the water is separate from the growing system, getting the right amount of water to the plants can be challenging. Dragging hoses and watering cans to and from the garden is onerous. Installed and automated irrigation systems address this, but their schedules don't take into account the needs of the plants or any environmental considerations. Whether the soil is hot and dry or soaked and muddy, the sprinklers go off at the same time for the same duration every week. This leads to over- and under-watering and can create plant stress, which reduces productivity and may make the plant more susceptible to pests and diseases.

In contrast, in an aquaponics environment, the grow beds are filled with water once an hour with a timer-based system and several times an hour with a siphon-based system. The result is a plant that "grows up" in a consistent water environment that never dries out. Because of the high levels of oxygen in the water and the oxygen that is pulled through the grow beds every time they drain, there is no chance of over-watering or drowning your plants. The plant develops a smaller root system in this predictable environment than it would in dirt because it never has to seek water and oxygen. With less energy directed toward root growth, the plant can focus its energy on growing foliage and fruit.

An even more important contrast between the use of water in aquaponics and in soil gardening is the amount of water consumed. Because aquaponics is a recirculating system, the only water that is consumed is either that used directly by the plants or that lost through evaporation. In contrast, most of the water used in soil gardening evaporates into the air during irrigation or seeps through to the groundwater tables. Aquaponic gardening typically uses less than a tenth of the water of soil gardening. Since agriculture is the single largest consumer of water in the world today, currently using 70 percent of the potable water in the United States, this is significant evidence for the relative sustainability of aquaponics.

Food

Botanists have long known that all plants require 16 micro- and macronutrients to fuel their metabolic functions. The first three elements — oxygen, hydrogen and carbon — are gathered from the air (oxygen, O_2, and carbon dioxide, CO_2) and water (H_2O). The remaining 13 elements are divided between macronutrients and micronutrients:

Macronutrients — calcium (Ca), nitrogen (N), magnesium (Mg), phosphorus (P), potassium (K) and sulfur (S)

Micronutrients — boron (B), copper (Cu), iron (Fe), manganese (Mn), molybdenum (Mo), zinc (Zn) and aluminum (Al)

In nature and in organic soil gardens, these nutrients are added to the soil through the decomposition of plant and animal matter. Leaves fall to the forest floor. Animals die and their bodies are returned to the prairie. Cow manure is gathered on the farm, aged and added to the compost pile. Worms and insects tunnel their way through the soil, leaving behind valuable waste products and oxygenated loam.

Once the plant and animal matter has been decomposed into a soil-like compost consistency, it then goes through a step called "mineralization" before becoming useable by the plants. Mineralization is the final breakdown of the matter by microbes into elements — nitrogen, potassium, etc. — that can then be taken up by the plants.

In hydroponic gardening, these steps are all bypassed and the nutrients are provided to the plants already in the mineralized form. Hydroponic nutrients are a careful mix of mineral salts that are formulated to optimize the plant's growth stage. Those salts are dissolved in water and added to water in a nutrient reservoir, where they are administered to the plant according to the style of hydroponic system being used. The nutrient level is frequently monitored using an EC (electrical conductivity) meter. Depending on the plant type, its stage of growth and environmental conditions, some nutrients are depleted from the nutrient solution over time. Others may become concentrated as they go unused and water is withdrawn. Water is added to the solution as it is used, but every two to three weeks the entire nutrient solution needs to be dumped out and replaced.

An aquaponics system feeds the plants in a way that combines the best of soil gardening and hydroponics. Because fish are cold-blooded and their waste is largely water soluble, it does not require composting. The waste water can be pumped directly into the planting beds where bacteria and worms mineralize what needs to be converted and made available to the plants. Even before the bacteria are established and the worms have been added, plants are able to take up some nitrogen directly through the ammonia the fish excrete. They prefer nitrate over ammonia, however, because nitrates require less energy to uptake and are easier for them to convert into protein internally.

Because the fish are constantly excreting waste, and that waste is constantly being digested and converted by the bacteria and worms, and because there are no salts to build up, there is no need to ever dump and replace the nutrient solution. As long as you follow the guidelines in this book for fish stocking density, grow bed to fish tank ratios, pH levels, etc., you will achieve a balanced state with your nutrients just as you would in a well-run organic garden.

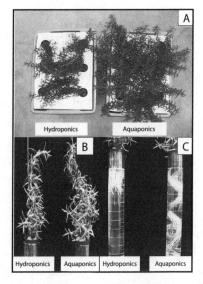

All of this won't happen overnight, however. The ability of an aquaponics system to convert fish waste into plant food is entirely based on the health and maturity of its bacteria and worm populations. When cycling is complete, Nitrosomonas and Nitrospria bacteria are present in your grow beds and are converting ammonia and nitrites into nitrates. Heterotrophic bacteria have also taken up residence in your grow beds to break down the solid waste from the fish, and hopefully you have added composting red worms to your beds. Over time, these bacteria and worm colonies become more established and your grow bed becomes increasingly more productive. Think of the environment in your grow bed as a fine wine that just keeps getting better with time.

Above: *Hydroponics vs Aquaponics rosemary.*

Images courtesy of Nick Savidov, PhD, Research Scientist, Leader of Greenhouse Crops Program, Crop Diversification Centre South Alberta Agriculture and Rural Development

After about six months of operation, an aquaponics system will be more productive than a similarly sized hydroponic system. A two-year study comparing aquaponics and hydroponics by Dr. Nick Savidov of the Crop Diversification Research Centre in Brooks, Alberta, Canada, showed not only

Bottom:

Hydroponics vs Aquaponics.

	Hydroponic			Aquaponic		
Crop Plant	**Height (cm)**	**Shoot (g)**	**Root (g)**	**Height (cm)**	**Shoot (g)**	**Root (g)**
Basil	30	226	68	35	301	111
Rosemary	31	141	119	35	226	290
Cucumber	138	1180	219	156	1580	274
Tomato	110	1616	198	114	1841	279

that that aquaponics overtook hydroponics in productivity after six months, but also that "The aquaponics system has an intrinsic capacity for self-regulation and balancing nutrients in the solution." (Savidov, 2005)

Because nutrient levels are relatively low in the beginning of your system's life, plants requiring relatively low levels of nutrients such as salad greens and herbs will do best. Save the tomatoes and cucumbers until after your system is at least a few months old and enough solids have gathered in the beds to provide the nutrients they need.

Air

Most of us learned in school that plants take up carbon dioxide and produce oxygen as a byproduct of photosynthesis. But did you know that plants also "breath" oxygen though their roots? This is why flooding a field or over-watering a potted plant will typically kill all its plant life — all of the air in the soil has been replaced with water. So how can plants grown purely in water, such as in hydroponics and aquaponics, survive? Because oxygen is in both the water and throughout the grow media. You are already driving as much oxygen as you can into the water in your fish tank for the health of your fish. The oxygen in that tank water is carried up and over the plant roots with each flooding cycle. Then, as the grow bed drains, more oxygen is pulled into the media behind the water.

Structure

Again, this may not have been in your earliest lessons on plants, but if you go back to the Media chapter in this book you may recall that one of the fundamental roles that soil plays in our environment is support. Have you ever noticed how plants that grow on the surface of a pond or lake tend to be very low and spread horizontally rather than stretch upwards? This is because a plant grown in the water without structural media for the plant roots to intertwine and grip on to would fall over very quickly. In soil gardens, the plant roots grow deeply to stand tall against the forces of gravity and wind. The same holds true in an aquaponic media bed.

Temperature

When considering the effect of temperature on plants, we typically think of the temperature of the air. The cool spring and fall temperatures are perfect for cool weather crops like lettuce, snap peas and spinach. The heat of the summer brings explosions of tomatoes, peppers and squash. In this respect, aquaponic gardening is no different from any other kind of gardening.

But what about the temperature where the roots live? In fact, root zone temperature plays more of a role in growth and development than that of the air surrounding the plant. Again, in the spring the soil will be cool and perfect for cool weather crops, but will be suboptimal, and possibly fatal, to warm-weather crops. Conversely, if you try to grow a lettuce plant in warm soil it will quickly "bolt" or try to reproduce itself by producing a flower stalk and become bitter. In fact, researchers have found that by maintaining an optimum root zone temperature, standard greenhouse air temperatures can be lowered 15 °F (8 °C) without affecting productivity. (Diver, 2002)

In aquaponics, the temperature of the root zone will be driven by the temperature of the fish tank water. If you are growing a warm-water fish such as tilapia, you will be able to maintain a root zone temperature that is compatible with most plants.

If you grow a cold-water fish such as trout, however, you will be limited to only growing cool season plants. Warm season plants can't survive root zone temperatures below 55 °F (13 °C), and won't begin to thrive until the temperature is above 65 °F (18 °C). Tomato plants, for example, prefer a root zone temperature of 75 °F (24 °C), which is happily the same temperature that tilapia thrive in.

Sunlight

We covered this topic in some depth in the System Location and Environment chapter, but for the purposes of this discussion, suffice it to say that the light requirements of soil-grown and aquaponically grown plants are identical.

How to start plants in, and for, aquaponics

Now that you know what kinds of plants you can grow in aquaponics, and how they will grow compared to soil and hydroponic gardening, let's discuss how to get those plants into your system.

Seeds

Starting plants from seeds is not only the most economical way to find plants for your aquaponics system; it is also the most fun. There are thousands of varieties of seeds available through your local nursery and online. Many might be familiar and bring back childhood memories but most will probably send you off on a new botanical or culinary adventure. Be bold!

Growing plants from seed can also have a positive global impact. We are experiencing a tragic loss of biodiversity through the effects of climate change and the patenting and commercializing of plant genetic material by megacorporations like Monsanto. A 2010 report released by the UN Food and Agriculture Organization estimated that 75 percent of crop diversity was lost between 1900 and 2000. This report also warned that this loss of biodiversity could have a huge impact on the ability of the world's population to feed itself as the global population rises to nine billion by 2050. "There are thousands of wild crop relatives that ... hold genetic secrets that enable them to resist heat, droughts, salinity, floods and pests," FAO director general Jacques Diouf said in the report. (AFP, 2010)

You can do your part to keep genetic diversity alive by purchasing heritage and heirloom varieties whenever possible. I have included a list of seed companies that specialize in these varieties in the Resources section of this book. Another great option is to learn how to save your own seeds from crop to crop. How to do this is not within the scope of this book, but it is not hard and there are many online resources and books available to guide you.

Broadcasting seeds

Clearly, the easiest way to start seeds in any environment — dirt or dirt-less — is through "broadcasting" or scattering the seeds evenly over the growing surface. Yes, this is even possible in a media-based aquaponic system. I have found that this technique works well for lettuce, radishes and carrots, and would probably work for other small seeds that are typically planted in the early spring and are adapted to being very wet. Simply scatter the seeds over the grow media. They will fall between the stones and many will reach the right depth to germinate.

Germinating seeds in a wet paper towel

The next category of seeds is those that still germinate very quickly, but are larger than lettuce. This includes beans, peas, melons and cucumbers. I have found that planting them directly into the grow beds usually doesn't work very well. They don't germinate as reliably as I know they can and because they are longer-term plants, I actually want to pay some attention to where they are positioned in the grow bed. But because they germinate and grow so quickly, it seems silly to waste much effort starting them. These are my candidates for the old first-grade trick of starting seeds in wet paper towels. Once the seeds are arranged, seal them in a large ziplock bag and check them daily for sprouting activity. Once you see a decent root (1-inch/25 mm or more), gently place them in the grow media to a level where the root will get wet in your flood cycles.

Seed starting using vermiculite, vermicompost, shredded coir or other soil-less medium

This is another terrific way to start seeds that require more care than the seeds described above.

- Start with a two-part seed-starting tray that consists of an insert and a watertight base tray. The most common configuration is a 10" x 20" (25 x 50 cm) 72-pack insert with a 22" x 10.5" (55 x 26 cm) tray. Place the insert inside the tray.
- Select your seed-starting medium. Be sure that the medium you select is relatively fine and will soak up (wick) water.
- Moisten the medium so it is damp, but not dripping.
- Fill each cup in the insert tray two-thirds full of moist medium.
- Place two to three seeds in each cup. Make sure you read the seed package for special instructions; some seeds may require a period of pre-chilling or soaking.
- Cover the seed with a thin layer of additional medium. Some seeds prefer a deeper planting depth than others. The rule of thumb here is to cover your seed with twice as much medium as the seed is wide. A small lettuce seed would therefore have a very fine cover layer and a large bean seed would have a much thicker cover layer.
- Add water to the bottom tray without disturbing the medium or the seeds in the inserts. Add enough water so it can soak up through the medium and reaching the seeds without soaking them. Be sure to use the nutrient-rich water from your fish tanks to give your seedlings a good start in life.
- Cover the tray with a humidity dome or plastic wrap to keep the humidity level high until the seeds germinate.

- Seeds for warm weather crops do best with some bottom heat. Either use a seedling heating mat or find a warm surface in your home, such as the top of a refrigerator or clothes dryer.
- Check the seeds daily and add water as needed.
- Once the seeds germinate, be sure to give them plenty of light (12–18 hours a day).
- Plant them in your aquaponics system once they have a couple sets of leaves and enough roots to intertwine through the seed-starting medium and be lifted from the cup.
- If more than one seedling is growing in the same cup, cut off all but the strongest seedling at the base. Don't try to pull out the extra seedlings, since this might damage the roots of the seedling you are keeping.

Seed starting using starter plugs

I sometimes use starter plugs when I'm just starting a few seeds and don't want to go through the hassle of preparing an entire seedling tray. But then the question becomes what starter plugs to use.

Rockwool — this is the default seed-starting plug for hydroponics. Pros — completely inert, so you will have a very sterile seedling with no chance of fungus or insects being harbored. Cons — needs to be pH balanced and is made of a spun rock material that is unpleasant to work with.

Peat sponges — I'm referring to brands such as Rapid Rooter®, Sun Leaves Super Starter and other similar products that are based on peat, latex and assorted "bio-yummies." Pros — no pH adjusting needed, they are largely biodegradable, grow well and are pleasant to work with. Cons — they have been known to be a breeding ground for fungal gnats and they can be expensive.

Plug seedling start.

Hydrofarm, Inc.

Cuttings

Cuttings are a great way to go if you already have access to plants. Cuttings root exceptionally well in aquaponics, especially tomatoes and peppers.

- To start plants from cuttings, you will want to make a diagonal cut through a young shoot from the mother plant, 3–5 inches (7.5 to 12.5 cm) long.
- Strip it of leaves, except for the leaves at the very top of the stem. The cutting needs some leaf growth to continue photosynthesis, since it can't take

in any food from roots it doesn't yet have. But too many leaves will just sap energy from its efforts to create new roots.

- Place the cutting on a flat, hard surface and make a clean slice through the middle of the stem with a sharp, sterilized blade.
- Insert the cutting directly into your media bed. Make sure that it is deep enough that water will come into contact with at least half the stem every time you flood your beds.
- Cover the cutting with a plastic bag until you see new growth. This reduces transpiration through the leaves and keeps the humidity level high while the cutting is establishing itself.
- If your cutting is in full sun, you will also need to provide it with shade until it has established itself.
- You will know that your cutting is rooted if you feel resistance when you gently tug on your cutting.
- Avoid using rooting hormones on your aquaponic cuttings as these can be harmful to the fish.

Buying plant starts

And finally, it is perfectly OK to have someone else start the plants for you. To ready a potted plant for your aquaponic system, remove the plant from the pot and shake off the excess dirt, then gently run the plant under water to further remove as much dirt as possible from the root system. Also, make sure you thoroughly check store-bought plants for bugs before planting.

Spacing your plants

Because aquaponically grown plants have plenty of everything they need at the root zone — water, oxygen and nutrients — each plant need grow only a minimal root system to supply its needs. Because resources are not scarce, the habitat is cooperative rather than competitive. The result is that plants can be placed significantly closer together than in a soil-based garden. In fact, most aquaponic gardeners space their plants about twice as close as they would in a dirt garden. The main consideration is what is happening above the "ground," not below it. Just be sure to space your plants so that light can penetrate the leaf canopy and air can circulate through and around the plants to help control insects and disease. If you don't get it right at first, don't worry. Because of the relatively small root balls in aquaponically grown plants, they are usually fairly easy to dig up and move around if you decide you need to give them more room as they mature.

Unhealthy plants

If your aquaponic system is fully cycled, and your plants don't look healthy, i.e., they aren't growing well and/or they have yellowed, curled or falling leaves, you need to figure out what is going on. The problem could be a number of things, but the most likely candidates are insects or nutrient deficiencies. Plants in aquaponics systems tend not to get diseases. In fact, the disease called pythium or "root rot," estimated to kill 30 percent of hydroponically grown plants, is virtually unknown in aquaponics. Researchers think that this is because, while hydroponics is a largely sterile system, aquaponics systems are full of beneficial bacteria and microbes that help plants to combat disease.

pH and nutrient supplementation

An aquaponics system that has been constructed, started and operated using the guidelines in this book should rarely, if ever, experience nutrient deficiencies. If you have the right amount of fish, and are feeding them a high-quality feed, and their waste is going into a properly sized, healthy biofilter, a near perfect, complete plant food will be the result. Here is the hitch. Plants can only take up certain nutrients within certain pH ranges.

In the chart on the following page, the thickness of the bar represents the amount of nutrient that is available at that pH level. Look at how much of each of the nutrient is available at the recommended pH level of 6.8–7.0, vs. a pH level of 6.0 or 8.0. Let's consider iron first, because it is the most common nutrient deficiency in aquaponics. Notice how the amount of iron that is available quickly decreases as pH rises above 7.0. This is called "nutrient lock-out."

When someone tells me that they have an iron deficiency I always go straight to pH. If it is below 7.0, then I want to know about the fish load in their tank and what they are using as food. I am a staunch believer that the only supplement a well-running aquaponics system should need is possibly calcium carbonate and potassium carbonate to buffer pH. As Dr. Wilson Lennard said to me once, "If you start relying on additives for your aquaponics system, you really just have a hydroponic system that grows fish."

strongly acid			medium acid	slightly acid	very slightly acid	very slightly alkaline	slightly alkaline	medium alkaline	strongly alkaline	
					nitrogen					
					phosphorus					
					potassium					
					sulphur					
					calcium					
					magnesium					
		iron								
		manganese								
		boron								
		copper & zinc								
					molybdenum					

4.0 4.5 5.0 5.5 6.0 6.5 7.0 7.5 8.0 8.5 9.0 9.5 10.0   *pH chart.*

I'll get down off my soapbox and say that there are a couple additives that are often used in aquaponics, especially in the first few months when the system is still young. The first is chelated iron, which is harmless to the fish and can quickly solve an iron deficiency problem that is unexplained by pH or fish output. Use it only sparingly, however, as it is very potent and a little goes a long way. The iron oxide produced by rusty nails is not a good source of iron in an aquaponics system because it is not available to the plants in a form them can use.

The other is a liquid seaweed product such as Maxicrop or Seasol (Australia). Many aquaponic gardeners add a small amount of liquid seaweed to their systems on a regular basis, almost as you would take a vitamin tablet as insurance against deficiencies in trace elements. Do you need to take a vitamin if you have a good diet? I wonder.

Insect control

Our final topic in this chapter is the little critters that like to eat plants even more than we do. Insects are, unfortunately, a part of a gardener's world. The good news is that you will most likely have far fewer insect problems in an aquaponic system than you would with a soil garden because many of the most harmful garden insects seek soil for their larval stage.

The best framework for thinking about insect pest control is integrated pest management (IPM). This is a four-step process that starts with set action thresholds. (US Environmental Protection Agency, n.d.) You will probably have different tolerances for how many insects you need to see before you take action, depending on where your aquaponic garden is set up. If it is a small garden attached to an aquarium in your dining room, I bet you will have far less tolerance for insects than if you have a large outdoor system. Because indoor and greenhouse environments don't have predator insects naturally available, harmful insect populations can also get out of control much more quickly than in outdoor setups. I tend to ignore insects in my outdoor gardens, and just remove the afflicted plant if it gets too bad. In the greenhouse, however, I have to take action quickly.

The second step in an IPM program is to monitor and identify pests. The best way to do this is through careful inspection of every plant that you harvest, and by occasionally walking around your garden and looking under random leaves. Also, hanging yellow and blue sticky traps will catch the flying insects and make it easier to not only identify them but also to see if their population is increasing or stable. Identify the insects using a magnifying glass or "loupe" and decide whether or not you are dealing with a harmful insect — not all insects are eating your plants, some may be eating other insects.

Hydrofarm, Inc.

Sticky trap.

The third step in an IPM program is prevention. In a greenhouse environment, this may mean sealing any cracks, installing insect screening on open windows and carefully inspecting any plant that enters the greenhouse. You should also keep your furry pets out of the greenhouse as they could be carrying hitchhiking pests (this is a case of do as I say, not as I do because my dog insists on always being with me in my greenhouse). If you are outdoors, you may consider hanging insect screening from a frame to reduce harmful insect pressure.

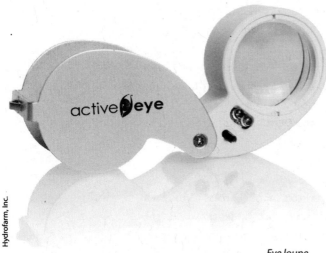

Hydrofarm, Inc.

Eye loupe.

The fourth and final step is control. Once monitoring, identification and action thresholds indicate that pest control is required, and preventive methods are no longer effective or available, it is time to plan a course of action for getting rid of the pests. This plan should start with what is least risky to the plants and fish, and move to the next method only if necessary. Here are some techniques I recommend, in order of least to most risky.

Feeding them to the fish

The best kind of bug problem is one that can be turned into fish food. Fish, and chickens if you have them, love slugs, caterpillars and other insect larvae. Just toss them into the tank and see what happens. In fact, if you get really bold, you may want to set some traps in order to actually attract them to your garden — free food! I know of aquaponic gardeners who have set up bug zappers over their fish tanks for this very purpose.

Even small insects, such as aphids, can be fed to the fish. If the infested plants are small enough, you can gently pull them out of the grow bed, shake off the loose media, and let them soak in the fish tank. I find that the aphids drown and release their grip on the plant in about fifteen minutes, and become fish food. If you get lucky the fish will even come up and nibble them off the plant. After the plant has been "cleaned up," just replant it back into the media. You may want to spray the surrounding media with an organic solution (see page 169) first, to kill whatever may have been shaken from the plant through your handling.

Knocking them off with a water spray

Because many harmful insects are at their most damaging stage when they are not very mobile, you might be able to control the population by simply knocking them off the plant with a stream of water. This is especially useful as an easy first step to controlling a large infestation that has gotten out of control.

Bringing in other insects to eat or otherwise kill them

There is an entire industry built up around supplying "beneficial insects" and it is now possible to find online companies that supply a wide variety of predators for the bugs that are bugging your plants.

Beneficial insects kill another insect in two ways: by eating it ("predators") or by laying their eggs within the host, ultimately killing it ("parasitoids").

Predators that like to feast on other bugs include species of beetles, lacewings, flies, midges, spiders, wasps and predatory mites. These insects tend to be larger than their prey, and not terribly selective about what bugs they eat.

The most commonly used predatory insect for pest control is the ladybug, also known as the lady beetle. Ladybugs are harvested in the foothills of California in the late summer and winter, and are widely available through garden supply shops, hydroponic stores and online. They love aphids and will consume more than five thousand aphids during their lifetime. After a few days of feeding, the female ladybug will deposit her eggs in small yellow clusters under a plant leaf or on the stem. The number of eggs laid depends on the pest population. In most cases, eggs are laid on or near plants infested with large numbers of aphids. Within seven days, the tiny eggs hatch into alligator-shaped larvae, which quickly begin feeding on many soft-bodied pests, mites and insect eggs. Within a month, the larvae will pupate and one week later, young adults will emerge, ready to feed. There are one to two generations per year, depending on weather conditions. (Shelton, n.d.)

The second most commonly used predatory insect is the lacewing. The beautiful green lacewing itself doesn't consume insects, but its larvae do, and have actually been nicknamed "aphid lions" for their prowess. Besides aphids, they feed on just about any soft-bodied pest they can "grab," including citrus mealybugs, cottony cushion scale, spider mites, thrips, caterpillars and insect eggs. They are known to be cannibalistic if no other food source is available. During the two to three weeks of the larval stage, they will devour up to two hundred victims a week. After this, they pupate by spinning a cocoon with silken thread; approximately five days later the adult emerges to complete the life cycle. There are five or six overlapping generations each season.

Parasitoids start their life on or within an insect host, ultimately killing it, hence their value in pest control. Most beneficial insect parasitoids are wasps or flies, although some rove beetles and other insects may have life stages where they are parasitoids. Most insect parasitoids only attack a particular life stage of their target insect species. The immature parasitoid develops on or within the pest, feeding on body fluids and organs and eventually leaving the host to pupate or emerging as an adult.

The most widely used parasitoid in greenhouse pest management is a tiny parasitic wasp called Encarsia formosa used to control whitefly. While I have successfully used ladybugs and lacewings to control aphids, using Encarsia formosa to eradicate whitefly from my infested greenhouse was the most impressive pest control treatment I've ever witnessed. This aggressive wasp is attracted to its host by the smell of honeydew produced by the whitefly, and is capable of detecting this from several feet away. Adult parasites not only feed on this honeydew, but also on whitefly body fluids through a hole made in the pest larvae. A major factor affecting whitefly control with Encarsia formosa is temperature. At temperatures below 62 °F (17 °C), the parasites will not fly and consequently will not seek out new hosts.

Pests attacked by parasitoids die more slowly than those attacked by predators. Some are paralyzed, while others may continue to feed or even lay eggs before succumbing to the attack. Parasitoids, however, often complete their life cycle much more quickly and increase their numbers much faster than many predators.

Because most beneficial insects develop wings at some stage in their lifetime, this solution is best for covered, greenhouse and indoor environments. In the open outdoors, you hopefully have a free supply of beneficial insects already. You can encourage helpful visitors by planting "insectary plants." Fennel, coriander, dill and cosmos are all considered good plants for attracting beneficial insects.

Spraying with organic solutions

While still effective and relatively harmless, this is my third favorite option mainly because I tend to avoid adding any foreign substance to an evolving aquaponic ecosystem. That said, if you carefully avoid spraying into your fish tanks ("overspray"), none of these should do any damage to your system.

The first spray is the least risky. It is a combination of molasses, water and dish soap that has been popularized by Murray Hallam in his *Aquaponics Secrets* video. This treatment is not intended to kill garden pests, but rather to sweeten the taste of the leaves, thus making them less attractive. In his video, Murray takes a liter-sized spray bottle, adds a few tablespoons of molasses, fills it up with water, then adds about a teaspoon of dish soap to help the solution stick to the plants. He then shakes it vigorously and sprays it on his plants. He repeats this treatment once every few weeks, and says that it has really helped his insect problem.

To actually kill insects using a spray, I recommend either an organic insecticidal soap or neem oil. Insecticidal soaps work on contact in two ways. First, they wash away the protective coat on the surface of the insect's body. Once inside, the soap will break the cell membranes and the cells will die. Insecticidal soaps are most effective on small, soft-bodied insects such as aphids, spider mites, thrips, whitefly and mealybug. (Penn State University, 2003) The downside of using them in aquaponics is that the soap is not good for the fish. When applying, be very cautious about over-spraying into the tank water or on the flooded surface of your grow bed.

Insecticidal soap can be purchased online or through your local garden center, or you can make your own. The internet is full of homemade recipes, so finding a source you trust is critical. Here is one from *Rodale's Organic Gardener's Handbook of Natural Insect and Disease Control:*

"To make a batch, combine 1 garlic bulb, 1 small onion, and 1 teaspoon of cayenne pepper in a food processor or blender and process into a paste. Mix into 1 quart of water and steep for 1 hour. Strain through a cheesecloth and add 1 tablespoon of liquid dish soap. Mix well. The mixture can be stored for up to 1 week in the refrigerator."

Neem oil works differently. Rather than being simple poisons, the ingredients in neem are similar to the hormones that insects produce. Insects take up the neem oil ingredients just like natural hormones; these enter their bodies and block the real hormones from working properly. As a result the insects "forget" to eat or mate, or stop laying eggs. Some forget that they can fly. Obviously insects that are too confused to eat or breed will not survive. The population plummets, and they disappear. The cycle is eventually broken. (Bradtke, n.d.)

Finally, there are a few commercial organic pesticides that are considered to be safe to use around fish. The first is "Dipel®," which consists of a powder made up of Bacillus thuringiensis, a naturally occurring bacterium that specifically attacks the caterpillars of moths and butterflies. Mix with water and spray onto heavily infested vegetation. "BotanicGard®" (Beauveria bassiana) is a similar product developed for controlling small insects. Both of these organic pesticides rely on microorganisms to attack specific bugs and are inert to everything else. They are approved for use up to the day of consumption as the active ingredients only harm bugs, not humans or fish.

Plants — conclusion

You may be coming to aquaponics as an experienced gardener or a novice, or may have a persistent brown thumb. No matter what your past is I think you will find growing plants in an aquaponics system to be an entirely unique experience. No weeds, no watering, no fertilizing. Only incredibly growth from your aquaponic planting beds. What's not to love about that?

Aquaponic Plants Rules of Thumb

- Avoid plants that prefer an acidic or basic soil environment. Otherwise just about any plant can be grown in an aquaponics system.
- Plants can be started for aquaponics the same way they would for a soil garden — by seed, cuttings or transplant.
- If your plants are looking unhealthy after the first few months it is probably for one of two reasons:

  - Nutrient imbalance caused by out-of-range pH — Maintain pH between 6.8 and 7.0 for optimal nutrient uptake by your plants.
  - Insect pressure.

13

Bacteria and worms

"Aquaponics is simply the answer. As humankind ages, it faces its greatest challenge in securing the one most essential element, food. Without it, humankind will cease to exist. Aquaponics is the solution for the ever-growing lack of available natural resources necessary to sustain life."

— Gina Cavaliero and Tonya Penick,
Green Acre Organics, Brooksville, Florida

The real magic of aquaponics happens deep inside our grow beds. According to Dr. Wilson Lennard, there is actually more weight in the biomass of the microorganisms of an aquaponics system than in fish. Within that wet, dark slimy world, the ammonia, liquid waste and solid waste from the fish are constantly being converted to plant food by the bacteria and worms that live there.

Bacteria farmers

"Grass farmers grow animals — for meat, eggs, milk, and wool — but regard them as part of a food chain in which grass is the keystone species; the nexus between the solar energy that powers every food chain and the animals we eat," wrote Michael Pollen in a description of the farming philosophy of Joel Salatin of Polyface Farms in his book *The Omnivore's Dilemma* (Pollen, 2006, 188). In other words, by focusing on growing and managing the health of his fields, the grass farmer "uses" his animals as inputs into his farming system and not as the end result. They provide the fertilizer that stokes the engine of decomposition in the grassy plains that in turn explodes in nutritious grass that feeds the animals.

Will Allen, the founder and CEO of Growing Power in Milwaukee, calls himself a soil farmer. He creates alchemy with grocery store refuse, brewery waste and worms — lots and lots of worms. If Salatin's animal workers produce meat and eggs, Allen's slithering friends produce worm castings that in turn produce the salad greens and watercress that he sells throughout Milwaukee. At Will Allen's Growing Power, the mantra is, "If you produce good soil, the rest will follow." The nexus for Growing Power between trash and organic produce are their worms.

Both of these luminaries of the sustainable agriculture movement have learned to focus their farming attention on the nexus, or the means of connection, between the inputs and the result. The manure and sunlight are connected by the grass to the meat; the garbage is connected to the soil by the worms. By nurturing this relationship, each partners with the power of the earth to achieve a more productive result than if they had merely tried to harness and control the inputs and the outputs.

Following this logical thread, we aquaponic gardeners are proud bacteria farmers. The animals that fuel our systems are our fish, but the engine of the system is the nitrifying bacteria that convert the fish waste from poisonous ammonia into nurturing nitrates. Without a healthy colony of bacteria, the plants would have no food and the fish would die within days or even hours from their own toxic waste. The biofilter of the bacteria clinging to the media bed ages like a fine wine. Bacteria should be cultivated, cared for, prized and shared with other aquapons. Treat them well, respect them and they will reward you mightily.

Nitrifying bacteria

While there are literally hundreds of types of bacteria that will come to call your aquaponics system "home," there are two on which we primarily focus because they do the job of converting the toxic ammonia from the fish into benign nitrate: *nitrosomonas* and *nitrospira*. They are of a family of autotrophic bacteria called "nitrifying bacteria." The central focus of the cycling process when we start up our aquaponics systems is attracting these two naturally occurring friends to our systems.

A few points that even the non-scientist might find interesting about nitrifying bacteria. (Industries, n.d.)

- They are "aerobic autotrophs." This means that they require air to survive, use inorganic compounds (ammonia and nitrites) as an energy source and generally cannot use organic materials. The other bacteria that play a significant role in your aquaponics system are heterotrophic bacteria, which break down organic material.
- Nitrifying bacteria need a surface (gravel, Hydroton®, the walls of your plumbing, even plant roots) where they use the sticky slime they secrete to attach themselves and colonize.
- They are extremely efficient at converting ammonia and nitrites. Studies have shown that Nitrosomonas bacteria are so efficient that a single cell can convert ammonia at a rate that it would take up to a million heterotrophs to equal.
- Nitrifying bacteria reproduce through binary division, and will double every 15–20 hours, depending on conditions. While this may seem fast, when compared to heterotrophic bacteria that can double in as little as 20 minutes, it is actually relatively slow.
- In the time it takes a single Nitrosomonas cell to double, a single heterotrophic E. coli cell would have produced a population exceeding 35 trillion cells.
- Another difference from heterotrophic bacteria is that nitrifying bacteria must be in a moist environment in order to survive.
- Until recent advances in culture-independent molecular methods, scientists thought that the bacteria responsible for converting nitrites into nitrates was the genus Nitrobacter instead of what is now known to be *Nitrospira sp.*

Nitrosomonas bacteria create nitrites as a byproduct of their consumption of ammonia. Those nitrites, which are still toxic to the fish and a poor source of nitrogen for the plants, attract the second type of nitrifying bacteria: Nitrospira. Nitrospira consume the nitrites and give off nitrates. Nitrates are largely harmless to your fish and an excellent source of nitrogen for your plants.

Though I am not as well equipped as I would like to give a more complete, science-based explanation for this "magic," Dr. Lennard pointed me to these simple formulae that explain the chemical process that happens when the nitrifying bacteria go to work:

Nitrosomonas:

$$NH_4^+ + 1.5\ O_2 \longrightarrow NO_2^- + 2\ H^+ + H_2O + 84\ kcal/mole\ of\ ammonia \quad (1)$$

Nitrospira:

$$NO_2^- + 0.5\ O_2 \longrightarrow NO_3\text{-} + 17.8\ kcal/mole\ of\ nitrite \quad (2)$$

Overall:

$$NH_4^+ + 2\ O_2 \longrightarrow NO_3^- + 2\ H^+ + H_2O + energy \quad (3)$$

Total hydrogen ion (H^+) production from the entire process of nitrification, the conversion of one mole of NH_4^+ to one mole of NO_3^-, is 2 moles of H^+. (Eberling, 2000)

Caring for and feeding bacteria

Like plants and fish, bacteria require food, oxygen and the right environmental conditions in order to survive and thrive. In their case, ammonia is the start of the food chain that leads to the production of nitrates.

Bill's aquaponics story

I live high in the Rocky Mountains and have very little luck growing anything outdoors due to the short growing season. Nighttime temps go below 40 degrees [4 ºC] even in June and winter can start in early September. Also, here we have a nice variety of plentiful animals — from deer to chipmunks and mice, which will eat about anything and they can get through just about any kind of screening. They love almost any kind of leaf so just about anything that grows above ground is fair game.

Somewhat out of desperation, I started an aquaponics operation in my woodworking shop this past winter. I used home-made containers, two 55-gallon [210-l] aquariums and koi and goldfish that I already was raising as a hobby and put them together with some pumps and tubing to circulate the aquarium water to the plants and back.

Thus far this operation has shown far more promise than anything else I have tried. Systems based on soil simply do not work very well in this environment and climate. In addition to the pests and climate, the soil is poor (even with a lot of enrichment from compost) and it is almost impossible to keep a garden area moist when the wind blows — which is a great deal of the time.

I am further impressed with these features and the ecologically sensitive nature of an aquaponics operation. I am using approximately ☞

The population of bacteria ultimately relies on the availability of ammonia. When you have few fish, or they are not feeding well, the ammonia level will decline and the bacterial population will decline as well. This could cause problems if, for example, you have an outside system during a rapid transition from the cold of winter to very warm spring days. If the fish have not been feeding due to the low water temperatures, then the population of bacteria will drop off. If the weather rapidly changes and the fish suddenly switch to feeding mode, the rise in ammonia levels may overwhelm the bacteria until they can recolonize. If you can avoid these sudden temperature transitions, your system will stay much more in balance.

Temperature also affects the health and reproduction of nitrifying bacteria. The optimal temperature for bacteria reproduction is between 77–86 °F (25–30 °C). At 64 °F (18 °C) their growth rates is decreased by 50 percent. At 46–50 °F (8–10 °C) it decreases by 75 percent, and it stops altogether at 39 °F (4 °C). They will die off at or below 32 °F (0 °C) and at or above 120 °F (49 °C). (Industries, n.d.)

150 gallons [570 l] of water in the current setup and need to replace about 4 gallons [15 l] a week or less than 3 percent. This is a far cry from the amount of water that sinks through the ground and evaporates into the wind in a conventional garden or farming operation!

In two seedling planters that sit in or on top of the aquariums (thus receiving moist heat from the water), many seeds sprout in just a few days, even those that are labeled as needing 5 to 7, or even 10, days to sprout. This operation is too new to report on maturing and producing stages of development, but other sources claim that Aquaponics produces up to TEN times the amount of organic produce the same space the ground would and plants typically grow in half to one-third the time required for plants grown by conventional methods.

My results are not yet in, but I have a good and early start on growing some usable foodstuffs by spring and it is almost a sure thing that I will not be losing plants to the ground squirrels and rabbits. The plants are also protected from freezing in the heated shop and all of this is pretty low maintenance. When the ice melts on nearby lakes and streams I will add some trout to the fish tanks and perhaps we can be somewhat self-sufficient, at least for the summer!

— Bill Hahn, Allenspark Park, Colorado

Because the nitrifying and heterotrophic bacteria that are important to converting fish waste into plant food are aerobic, oxygen is as important to their health as it is to the fish and the plants. Nitrifying bacteria will die quickly if oxygen is cut off to the grow bed, for example. Even more alarming, heterotrophic bacteria will actually start converting the nitrates back into ammonia and nitrites in a low-oxygen situation (dissolved oxygen of less than 2.0 ppm) in a process called "dissimilation." This is why an area of your grow bed that becomes so clogged with solid waste that oxygenated water can't flow through it will become anaerobic and will smell badly. The smell is the ammonia gas being released from dissimilation.

Finally, bacteria require a pH between 6.0 and 8.5* in order to metabolize and reproduce. The optimum pH range for Nitrosomonas bacteria is 7.8–8.0. The optimum pH range for Nitrospira bacteria is between 7.3–7.5. At pH levels below 7.0, Nitrosomonas will grow more slowly and increases in ammonia may become evident. All nitrification is inhibited if the pH drops to 6.0 or less. (Hill, 2008)

Besides these factors, the other way to kill your bacteria is through harmful chemicals. Again, just like with all the living elements of your aquaponics system, use extreme caution when selecting insect control and dechlorinate the water you add to your tank.

Worms

We have been focusing on nitrifying bacteria, but another bacterium also plays an important role in your aquaponics system. These bacteria are heterotrophs, which in simple terms means that they consume organic carbon and give off energy (nutrients) for plants. The problem is that they generally can't keep up with breaking down the mass of the solid waste. Consequently early aquaponic gardeners were faced with two choices. First, they could leave the solid waste in their grow beds to take advantage of the extra nutrient it provided their plants, but they would need to regularly clean it out of their grow beds. The second option was to filter the solid waste before it entered the grow bed. Neither was very attractive.

Then a few years ago someone got the brilliant idea to add composting red worms to their aquaponic media bed. Worms are one of nature's great garbage disposals. As Murray Hallam is fond of saying, they are aquaponics "secret ingredient." Not only do they break down and digest the solid waste and dead root matter that plants slough off, but in return they give us another one of nature's perfect fertilizers. The waste from worms is called worm castings or vermicompost, and when it is steeped in water it becomes an incredibly potent fertilizer called "vermicompost tea" or "worm tea."

Researchers at The Soil Ecology Laboratory at Ohio State University have found that worm castings soaked in oxygenated water ("tea") provides the following benefits in to increasing seed germination and plant growth: (Arancon, 2007)

- Suppresses plant disease (including Pythium, Rhizoctonia, Plectosporium, and Verticillium)
- Suppresses plant parasitic nematode
- Suppresses plant insect pest (including tomato hornworms, mealy bugs, spider mites and aphids)

Said simply, worm tea increases the productivity of the plant side of your aquaponics system!

Whenever I talk about adding compost worms, or red worms, into aquaponic grow beds, someone will ask "but don't they drown?" It is a reasonable question. The answer is "no," they actually survive, thrive and multiply extremely well in an aquaponic grow bed. The secret again is the oxygen level. Because the water that is flooding the grow beds is full of oxygen, the worms, which actually breathe through their skin, are fine. Moisture also plays a key role in the worm's ability to breathe, and moisture is something that an aquaponic grow bed has plenty of!

There is no rule of thumb when it comes to how many red worms to add to your aquaponic media beds. Once you add the worms they will start to multiply. They will self-regulate their population and if they become overcrowded or don't have enough food they will stop reproducing. It is hard to have too many worms in your grow beds.

That said, it is possible to have too much of a good thing if you are adding worms, vermicompost AND worm tea to your system. According to Dr. Lennard, "Many people now add 'worm tea' from a worm farm that is not part of the aquaponic system [i.e., a separate worm farm]. Many people use vermicompost to raise their seedlings in. I saw an aquaponics system once that had fish, used worm tea additions, used vermicompost in the seedlings as media and added bacterial teas from bacterial digestors and lactose.

The system became so nutrient dense that by the time I saw it and asked about plant growth rates, they were growing at half the expected rate for that climate! You can overfertilize aquaponic systems if you willy nilly add nutrients."

The lesson there is to just add the worms and let them make the worm tea as part of your ecosystem environment. At a minimum level I recommend about a pound of worms for every 20 cubic feet (0.5 m³) of grow bed volume. You can often find a local source for red worms, including fishing bait shops. If not, there are many sources on the internet.

Your worms will probably arrive in a bag full of soil. The best way to separate the worms from the soil is to take advantage of the worm's natural aversion to light. Expose the bag with the worms to the sunlight or shine a bright light over the top. The worms will burrow into the soil to escape the light, freeing the top layer of any worms. Scrape off that layer and repeat the exercise. Eventually you will have removed most of the soil and can let the worms finally escape into their new home in your grow bed. There is no need to bury them; they will happily find their own way into the dark, slimy depths.

Bacteria and worms — conclusion

Fish and plants get most of the attention in an aquaponic system because they are visible, fun and are, after all, what most growers are working to produce. But it is inside the grow bed itself that the true magic of aquaponics happens. Without the transformation of the fish waste into plant food that the bacteria and worms cause, the fish would quickly die in their waste and the plants would soon follow from lack of food.

Aquaponic Worms Rule of Thumb

- Add a handful of composting red worms to each grow bed once your system is fully cycled and fish have been added.

Section 5

The integrated system

14

Cycling

"I am now in a wheelchair and I am working hard to find ways to garden that people with disabilities can do. I am convinced aquaponics is the answer."

— Jeffrey Mays, Richmond, Virginia

Now your system is fully set up, the water has been added and the plumbing works. It's time to add the fish and the plants, right? Not so fast! First we need to build the bridge between them by establishing the bacteria colony.

In Chapter 13: Bacteria and Worms, we talked about the critical role that bacteria play in an aquaponics system. They are the magic ingredient that takes the unusable fish waste and creates a near perfect plant fertilizer. Bacteria are what marry aquaculture and hydroponics into a symbiotic ecosystem without the flaws inherent in those two methods.

In this chapter we will demystify the process of establishing a beneficial nitrifying bacteria colony in your aquaponics system. This process is often called system "cycling" because it is the initiation of the nitrogen cycle in your system. By the end of this chapter, you will fully understand what you MUST do to initiate cycling and ensure its continued success. You will also understand what you CAN do to both make the process less stressful for your fish and your plants, and to speed it up.

What is cycling?

Cycling is really shorthand for establishing a biofilter where the nitrogen cycle can take place within your system. The nitrogen cycle is the ongoing

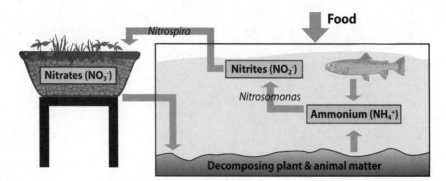

Bacteria cycling.

process in which bacteria convert ammonia and nitrites into food that your plants can consume. An aquaponics system is fully cycled when no, or almost no, measurable ammonia or nitrites are present.

Cycling starts when you (or your fish) first add ammonia to your system. Ammonia (chemical formula NH_3) is a compound made of nitrogen and hydrogen. It can come either from your fish or from other sources. Ammonia is toxic to fish and will quickly kill them unless it is either diluted to a non-toxic level or converted into a less toxic form of nitrogen. Unfortunately, nitrogen as found in ammonia is not readily taken up by plants, so no matter how high the ammonia levels get in your fish tank, your plants will not be getting much nutrition from it.

The good news is that ammonia attracts Nitrosomonas, the first of the two nitrifying bacteria that will populate your system. The Nitrosomonas convert the ammonia into nitrites (NO_2). This is a necessary step in the cycling process; however, the bad news is that nitrites are even more toxic to fish than is ammonia! But there is good news because the presence of nitrites attracts the bacteria we are truly after — Nitrospira. Nitrospira convert the nitrites into nitrates, which are generally harmless to fish and important food for your plants.

Once you detect nitrates in your water and the ammonia and nitrite concentrations have both dropped to 0.5 ppm or lower, your system will be fully cycled and aquaponics will have officially begun!

The importance of testing tools

You must have some way of telling where you are in the cycling process — typically a four- to six-week endeavor. Specifically, you must monitor ammonia, nitrite and nitrate levels as well as pH so that you know that all these elements are "in range." If they are not, you may need to take corrective action. This is also the only way you will know when you are fully cycled and ready to add more fish (or your first fish, if you have been cycling with no fish at all). Plus, watching the daily progress of the cycling process is fascinating and something you can only see through the lens of a test kit. By the way, once you reach the point that your system is fully cycled, you will need to do much less monitoring than during the cycling process. So get through the cycling process and look forward to reaping the fruits (… or should we say, the fish) of your labor.

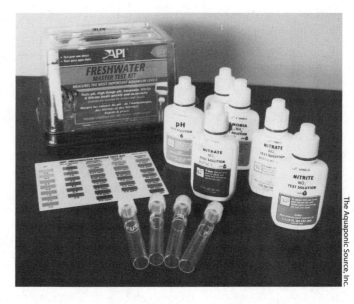

The Aquaponic Source, Inc.

API Freshwater Master Aquarium Test Kit.

Please note that I've included a worksheet in the appendices of this book to assist you in collecting the data you need to manage the cycling process. It is also available online at AquaponicGardening.com.

To do their testing, most aquaponic gardeners use a product by Aquarium Pharmaceuticals Inc. called the API Freshwater Master Test Kit. This kit is easy to use, inexpensive and is designed for monitoring the cycling process in fish systems.

You will also need a submersible thermometer to measure your water temperature. Temperature affects both the cycling rate and the health of your fish and plants once you are up and running.

The last piece of testing equipment you may want to have is a dissolved oxygen (DO) test kit or meter. Dissolved oxygen will speed up the cycling process because the more aerobic your environment the more attractive it will be to the bacteria. More importantly, dissolved oxygen is a critical factor in the health of both your fish and your plants. I consider this a "nice-to-have" rather than a "need-to-have," however, because if your system is being oxygenated as instructed in this book (specifically in the Water and Plumbing chapters), and your fish are eating well and not gasping for air at the surface of your tank, you probably have enough oxygen to cycle. That said, if you are planning on setting up a large system with a lot of fish you may want to have the extra degree of control that comes from knowing what your dissolved oxygen levels are. Be aware though — dissolved oxygen is a complex parameter to measure and you get what you pay for in testing equipment. In my experience, the inexpensive dropper-style kits don't work very well. You should spend the extra money on a digital meter if you decide to measure oxygen.

Initiating cycling

Ammonia (NH_3) starts the cycling process. You must have some means to release ammonia into the system so that you attract the bacteria that are at the heart of aquaponics. There are two ways to introduce ammonia into your system — with fish and without (fishless).

Cycling with fish

I used fish to cycle my first aquaponics system and I suspect this is how most people approach cycling. In some ways it is the easier of the two methods because there are no extra inputs, but it is definitely the more stressful of the two options because live critters are involved.

The idea is to add fish on day one and hope that they live through the cycling process. The challenge is to get the system cycled fast enough that the ammonia concentration from the fish waste drops to a non-toxic level before the fish succumb to exposure to their own waste. I strongly recommend that you don't stock to your tank's mature capacity (1 lb of fish per 5 gallons of water / 1 kg of fish to 10 l of water) but to less than half that. You might also want to consider these fish as "sacrificial" and perhaps use goldfish from the pet store, which are fairly tolerant of ammonia, rather than the tilapia with which you may ultimately envision stocking your tank. Also, do not feed these fish more than once a day and only feed them a small amount even then.

The ammonia (NH_3)

Fish excrete ammonia through their gills as a byproduct of their respiratory process. Without dilution, removal or conversion to a less toxic form of nitrogen, the ammonia will build up in the fish tank and eventually kill the fish. In addition, ammonia (NH_3) continually changes to ammonium (NH_4^+) and vice versa, with the relative concentrations of each depending

Percentage of toxic ammonia (NH$_3$) in a fish culture system at different pH and temperature levels								
temp (°F)	61	64	68	72	75	79	82	86	90
temp (°C)	16	18	20	22	24	26	28	30	32
pH									
7.0	0.30	0.34	0.40	0.46	0.52	0.60	0.70	0.81	0.95
7.4	0.74	0.86	0.99	1.14	1.30	1.50	1.73	2.00	2.36
7.8	1.84	2.12	2.45	2.80	3.21	3.68	4.24	4.88	5.72
8.2	4.49	5.16	5.94	6.76	7.68	8.75	10.00	11.41	13.22
8.6	10.56	12.03	13.68	15.40	17.28	19.42	21.83	24.45	27.68
9.0	22.87	25.57	28.47	31.37	34.42	37.71	41.23	44.84	49.02

VanGorder, 2000, 8

on the water's temperature and pH. Ammonia is extremely toxic to fish; ammonium is relatively harmless. At higher temperatures and pH, more of the ammonia is in the toxic form than at lower levels.

Standard test kits measure total ammonia (ammonia plus ammonium) without distinguishing between the two forms. The above chart gives the maximum long-term level of ammonia in ppm that can be considered safe at a given temperature and pH.

You will need to monitor your tank water daily during cycling for elevated ammonia levels. If your level exceeds the levels on the chart above, you should dilute with a water exchange by removing up to a third of your tank's water and replacing it with fresh, dechlorinated water. Be careful, however, that the new water is the same pH and temperature as the water remaining in the tank so you don't further stress the fish with rapid changes.

Adjusting pH

During cycling, especially with fish, you should try and keep your pH close to 6.8–7.0. The preferred range does not go below 6.8 because the bacteria will reproduce best between 7.0 and 8.0. But much above 7.0 and the ammonia toxicity issue described above can become problematic (higher pH readings suggest higher ammonia concentrations). So how do you keep pH in such a tight range?

The first rule is: Whatever you do to adjust pH when fish are present in aquaponics, do it slowly! Fast, large pH swings are very stressful on fish and will be much more of a problem than having pH that is out of range. Target shifting your pH reading no more than 0.2 per day and you should be fine.

We covered pH extensively in the Water chapter. Refer to that section if you need to adjust the pH in your system.

Typically, you will be trying to keep pH down during cycling. Then once your system is cycled you will probably notice that the pH will fall and you will need to switch to keeping it up. You will probably find that it is easier to increase pH than it is to decrease it. The ideal pH of a mature aquaponics system is also 6.8–7.0. This is a compromise between what the plants prefer, i.e., a slightly acidic environment of 5.5–6.5, and what most fish prefer, i.e., a slightly alkaline environment. This is covered in more detail in Chapter 15: System Maintenance.

Nitrite (NO_2)

Nitrite is to fish what carbon monoxide is to air-breathers. It will bind with the blood in place of oxygen and keep fish from getting the oxygen they need. Fish poisoned with nitrites die of what is called "brown blood disease." If the nitrite levels in your tank rise above 10 ppm while you are cycling your system with fish, you should do a water exchange as discussed above, but you may also want to add salt to further protect your fish's health.

Do this by adding salt to the system to at least one part per thousand (1 kg of salt per 1000 liters of water) using non-iodized salt. Avoid table salt since it contains potentially harmful anti-caking agents. You can use cheap pool salt or water softener salt available at the grocery or hardware store. Make sure to dissolve it completely in a bucket of water before adding it to the fish tank since undissolved salt crystals on the bottom of a fish tank can burn a fish that rests against them. Stop feeding your fish until the nitrite levels drop below 1.0 and aerate as much as possible.

Why does salt (sodium chloride) help? Sodium chloride helps mitigate nitrites because the chloride ions bind with the nitrites and thereby help keep some of the nitrite out of your fish.

Cycling with fish — conclusion

While cycling with fish is perhaps the most straightforward of the cycling techniques, I'm not a big fan. It is very stressful on your fish because they are being subjected to unnecessarily high levels of ammonia. They will not be feeling well, and may not come through the process alive. All of this is stressful on you as well because you will be worried about your fish.

Fishless cycling

The other way of cycling is to add ammonia by some means other than fish. This technique has a few major advantages. First, there is much less stress involved for you and the fish because you are not trying to keep anybody alive during the process. Because of this, you need be much less concerned about pH since the pH must only be kept in a range that facilitates cycling without consideration for the safety of the fish.

Second, because you can elevate the ammonia concentration to a much higher level than would be safe for fish, you can cycle your system in much less time (generally ten days to three weeks versus four to six weeks when you cycle with fish) and end up with a more robust bacteria base once you are cycled. The practical result of this is that you can fully stock your tank once cycling is complete, versus gradually increasing the stocking levels as is recommended when cycling with fish. This is especially beneficial to those who are growing aggressive or carnivorous fish because they are less likely to attack each other if everyone is introduced to the tank at the same time.

Finally, you can more precisely control how much ammonia is added to your system during the process. For example, if you see that your ammonia level is creeping up to eight ppm, but no nitrites have shown up yet, just stop adding ammonia for a few days and let the bacteria catch up. You can't do this with fish!

There are several ways to add ammonia to your system, ranging from the obvious to the slightly bizarre. I will talk about the pros and cons of each, and you can decide for yourself which makes the most sense to you.

Liquid ammonia (aka clear ammonia, pure ammonia, 100 percent ammonia or pure ammonium hydroxide)

This is the old-fashioned cleaning product your grandmother used to use that filled the room with the smell of ammonia. Only use it if you can find the pure form that is made strictly out of ammonia (usually 5–10 percent by weight) and water. Avoid anything with perfumes, colorants, soaps, surfactants or any other additives. Shake the bottle. If it foams or if it doesn't list the ingredients or says "Clear Ammonia," "Pure Ammonia," "100 percent Ammonia" or "Pure Ammonium Hydroxide," leave it on the shelf.

The hardest part of cycling with pure ammonia can sometimes be finding the ammonia. Where I live, I can buy it at our locally owned McGuckins Hardware. Try your local hardware store, cleaning supply store or even well-stocked superstores. If all else fails, you can order it online.

Pros — It is relatively inexpensive (approximately $10 for a gallon) and what you don't use to cycle your aquaponics system can be used to clean your windows! Plus, you know exactly what you are adding to your system with this product — ammonia and water, nothing more, nothing less.

Cons — It can be hard to find if you don't have a cleaning supply or a good hardware store nearby. I'm told that it is entirely unavailable in Australia since September 11, 2001, because of the remote association as a possible ingredient in bomb-making.

Ammonium chloride (crystallized ammonia)

This is the same concept as the liquid ammonia above, but you can find this through aquarium supply stores, soap supply stores, photography supply stores and chemical houses.

Pros — Because it is very concentrated and in dry form, it is inexpensive to ship. If you get the kind intended for aquariums, there will be little doubt that it is pure and will work in cycling.

Cons — There will be cost involved.

Human urine (aka "humonia" or "pee-ponics")

Sound gross? Well, when you think about it, human urine is actually an excellent source of ammonia, just as the waste product from any animal would be. Human urine is just easier to capture. Here is the catch. In order to go from urea to ammonia, you should put the urine into a sealed bottle for a few weeks to "percolate." Can you just urinate straight into the fish tank? Sure, but the problem is that since that urine will take a while to convert into ammonia, you will have no way of telling just how much potential ammonia you have in there. The levels will read very low, and then all of a sudden one day they will explode.

Pros — This is a free and readily available source of ammonia.

Cons — There is the "yuck" factor, the fact that you have to store the urine until it converts to ammonia, and the possibility that harmful bacteria or germs from your digestive system are transmitted to your aquaponics system.

Other sources of ammonia

As animal flesh decays it gives off ammonia. I saw a suggestion on a forum once for cycling your system using a bit of dead fish, but dismissed this concept as too bizarre. Then I was testing one of my fully established, rock solid tilapia systems with a group of people who had just taken a class from me and were learning about maintaining their system. Imagine my embarrassment when I found the ammonia reading was off the chart! Ends up that a fish had died in the back corner of the tank and hadn't floated to the surface.

Pros — This is another free and readily available source of ammonia.

Cons — Again, because other bacteria and chemical compounds are given off during the decay process, there is a chance that you will introduce something undesirable to your aquaponics system, not to mention the chance of attracting flies or other insects that want to assist in the decomposition. The other issue is that you will find it difficult to control how much ammonia gets into your system with this method.

Instructions for fishless cycling

Once you have identified your source of ammonia you are ready to start the cycling process. Just follow these simple instructions.

- Add the ammonia to the tank a little at a time until you obtain a reading from your ammonia kit of 2–4 ppm.
- Record the amount of ammonia that this took, and then add that amount daily until the nitrite appears (at least 0.5 ppm). Test daily. If ammonia levels approaches 6 ppm, stop adding ammonia until the levels decline back down to 2–4 ppm.
- Once nitrites appear, cut back the daily dose of ammonia to half the original volume. If nitrite levels exceed 5 ppm stop adding ammonia until they decline to 2.
- Once nitrates appear (5–10 ppm), and both the ammonia and the nitrites have dropped to zero, you can add your fish. Stop adding ammonia.
- Again, you will probably want to photocopy the data collection worksheet I've included in the appendices of this book to manage the cycling process. It is also available online at AquaponicGardening.com.

Adding plants

I recommend adding plants to your new aquaponic system as soon as you start cycling. Plants can take up nitrogen in all stages of the cycling process to varying degrees, from ammonia, nitrites and nitrates, but they will be happiest when cycling is complete and the bacteria are fully established because so many more nutrients become available at this stage.

Plants focus on establishing their root systems when they are first transplanted into a new environment. You may initially see some signs of stress — yellowing and/or dropped leaves — and you will probably not see any new growth for a few weeks. This is fine. Adding plants to your system right away lets them go through the rooting process early on and readies them to start removing the nitrogen-based fish waste from your aquaponics system as soon as possible.

In Australia, many aquaponic gardeners use a liquid seaweed product called Seasol to ease the cycling process for the plants and to help them establish their roots. The North American equivalent of this product is called Maxicrop, which can be found in garden centers, hydroponic stores, and online in both liquid and dry form. Maxicrop is derived from Norwegian seaweed, and is advertised primarily as a growth stimulant, especially enhancing the development of plant roots. It is extremely effective at giving plants a "leg up" after being transplanted into your new aquaponics system, is absolutely harmless to the fish and probably beneficial for the bacteria.

While there are no hard and fast rules about how much Maxicrop to add during cycling, I recommend about a quart of the liquid product for every 250 gallons (1000 liters) of tank water. It will turn your water almost black but don't worry; this will clear up after a week or so.

The Murray Hallam cycling technique

My Australian friend Murray Hallam swears by a simple cycling technique that is a hybrid between cycling with fish and without. He has successfully cycled hundreds of aquaponic systems by instructing his customers to do the following:
• Add liquid seaweed to the system at the rate described above.
• Add plants.
• Wait for two weeks.
• Then add fish (at a low stocking density if possible).

Murray has developed this technique because he can't get synthetic ammonia in Australia. The liquid seaweed, however, has a small amount of ammonia that begins the cycling process, easing the ammonia load on the fish so they aren't as stressed as they would be if they were added to the system on Day One.

Speeding up the process

Cycling is in some sense akin to any hunting activity that uses a lure. We start by putting out the ammonia. This attracts the Nitrosomonas bacteria which in-turn produces nitrites. The nitrites attract the Nitrospira bacteria which produce the nitrates that are harmless to the fish and delicious to the plants. These two beneficial nitrifying bacteria are naturally present in the environment.

As I stated earlier, this process will take four to six weeks if done with fish, or as little as ten days to three weeks if done fishless. But what if you could speed that up significantly? What if instead of waiting for the bacteria to show up to the party, they actually are part of the party to begin with? You can do this by introducing nitrifying bacteria into your aquaponics system.

Adding bacteria

While there are many ways to do this, they all boil down to two basic strategies: use bacteria from an existing aquaculture or aquaponics operation or from a near-by pond or instead, purchase bacteria from a commercial source. Remember from the Bacteria and Worms chapter that nitrifying bacteria require a surface to adhere to. Just adding water from an aquarium won't be nearly as effective as inserting the filter sponge from an aquarium into your grow bed.

Good sources of beneficial bacteria from existing systems are ranked here, leading with the best:

- Media from an existing aquaponics system.
- Filter material (floss, sponge, biowheel, etc.) from an established, disease-free aquarium.
- Gravel from an established, disease-free tank. (Many local pet and aquarium stores will give this away if asked.).
- Other ornaments (driftwood, rocks, etc.) from an established aquarium.
- Squeezings from a filter sponge (any pet and aquarium store should be willing to do this...).
- Rocks from a backyard pond that have fish in it.

There are also a number of commercial bacterial supplements (Cycle, Stress-Zyme, Bacta-Pur, Proline, etc.) available. These are reported to have mixed results, however, and are almost always far more expensive than just finding an existing source of live bacteria.

Managing water temperature

Water temperature also has a dramatic effect on the speed of cycling. Their optimal temperature is 77–86 °F (25–30 °C). At 64 °F (18 °C) their growth rate is decreased by 50 percent. At 46–50 °F (8–10 °C) it decreases by 75 percent, and it stops altogether at 39 °F (4 °C). It will die off at or below 32 °F (0 °C) and at or above 120 °F (49 °C).

Cycling — conclusion

Cycling your aquaponics system can be a stressful, mystifying time. You just want to get going with a fully stocked tank of fish and incredible plant growth. Instead you are possibly faced with stressed fish, yellowing plants and a courtship dance with invisible bacteria. Be patient. Establishing an ecosystem takes time and your efforts will be well rewarded with a stable, productive system for as long as you garden.

Aquaponic Cycling Rules of Thumb

Photocopy the data collection worksheet in the appendices to manage the cycling process. It is also available online at AquaponicGardening.com.

Cycling with fish

- Add only half as many fish as you would to be fully stocked.
- Test daily for elevated ammonia and nitrite levels. If either gets too high do a partial water exchange.

- Feed once per day or less to control ammonia levels.

Fishless cycling

Sources of ammonia:
- Synthetic — pure ammonia and ammonium chloride.
- Organic — urine and animal flesh.

The process

- Add the ammonia to the tank a little at a time until you obtain a reading from your ammonia kit of 2–4 ppm.
- Record the amount of ammonia that this took, and then add that amount daily until the nitrite appears (at least 0.5 ppm). Test daily. If ammonia levels approaches 6 ppm, stop adding ammonia until the levels drop back down to 2–4 ppm
- Once nitrites appear, cut back the daily dose of ammonia to half the original volume. If nitrite levels exceed 5 ppm, stop adding ammonia until they drop to 2.
- Once nitrates appear (5–10 ppm), and both the ammonia and the nitrites have dropped to zero, you can add your fish. Stop adding ammonia.
- pH should be between 6.8 and 7.0
- Plant your system as soon as you begin cycling. Adding liquid seaweed will help your plants quickly acclimate to their new environment.
- Adding bacteria will dramatically speed up the cycling process. Keeping the temperature of your water above 70 °F (21 °C) will help as well.

The Murray Hallam cycling technique

- Add liquid seaweed to the system.
- Add plants.
- Wait for two weeks.
- Then add fish.

15

System maintenance

"Aquaponics is freedom from pulling weeds!
The part I hated most about gardening was the weeding,
and now that's done away with!"

— Jim Knott, Grass Range, Montana

One of the best aspects of aquaponic systems is that they require far less maintenance than any other growing system. Unlike traditional soil gardening, there is no weeding, no watering and no fertilizing. Unlike hydroponics you don't have to regularly dump out and replace the nutrients and aquaponic plants are far less prone to disease, so there is less monitoring required in general. Even aquarium and aquaculture systems require occasional water changes and filter cleaning.

But like anytime we depart from a totally natural system and create a man-made version, we need to step in occasionally to keep things running smoothly. What follows is a list of daily, weekly and monthly maintenance tasks for your aquaponic system. You certainly will have some flexibility about timing. The following gives you a general idea of what you need to do and roughly how often you will need to do it.

I've also included a Monthly Maintenance Checklist in the Appendices and on AquaponicGardening.com. Keep a photocopy of this list by your aquaponics system to help you remember and record what you need to do to keep your system healthy.

Daily

Feed the fish

You should feed your fish at least once a day, and preferably twice: once in the morning and once as the sun sets. Not only should you do this for the obvious nourishment reason, but almost as an important daily check on the health of your fish. This is the main reason why I am not a fan of automatic fish feeders, unless you are out of town for a few days. If the fish don't eat vigorously, something is wrong; see Troubleshooting for ideas on what that might be and how to solve it.

Check the fish tank temperature

This is a fast check that could have a big impact on the health of your fish; see the Fish chapter for details on what the acceptable temperature range is for the fish you are raising.

Check the pumps and plumbing system

After you have fed the fish and made sure the tank temperature is in your expected range, check the circulatory system, i.e., the pumps and plumbing.

If you are using a flood and drain system that is triggered by a timer, you need to make sure that both the pump and the timer are functioning properly. If your timer is on an "on" cycle and your pump is running, clearly both your timer and your pump are working. The only possible problem at this point is that your timer doesn't go to the "off" cycle. In my experience, this is pretty rare, but the evidence will be that your plants will start wilting from lack of oxygen.

If your pump is not on when you are checking your system, the best way to see if it has gone off lately is to run your finger tips against the inside wall of your media guard. If it is dry, you have a problem. The next step is to bypass the timer cycle by flipping on the override switch on the timer, or to force the timer to an "on" position. If the pump doesn't go on, you have a pump problem, and it could be both a timer and pump problem. If the pump does go on, you have a timer problem.

If you are using a siphon the process is very similar. If the pump is on, you know that it is OK. Stay until the siphon should trip (usually 7–10 minutes). If there is a problem, it will most likely be that the accumulation of muck in the pipes is slowing down the water flow enough to prevent the siphon action from starting. If it does not start, you are probably due for a thorough pipe cleaning.

Weekly (after cycling)

Check pH

pH is arguably the single most important determinant of the health of your aquaponics system. pH is what governs your plants' ability to take up nutrients, your bacteria's ability to reproduce and the vitality of your fish. Once your system is cycled and has been stable for a while, it is easy to get into a false sense of security with your pH and stop taking regular readings. Don't stop!

While some aquaponic systems will hold a steady pH level of 6.8–7.0 (mine does, for example), most healthy aquaponic systems will see their pH naturally decrease over time. Once you see pH decline to 6.4 you should take action to "buffer up" by adding hydrated lime or potash (see the pH section in the chapter on Water for more details).

If the pH starts rising, you are either adding something to your system that is causing it to rise (hard water, seashells in your media, etc.) or you have an anaerobic "dead zone" in your media beds somewhere. You should figure out what is causing this to happen and correct it, if possible.

Check ammonia levels

Next to pH, ammonia levels are probably the most important indicator of the overall health of your system. Even though you are fully cycled and your system seems to be functioning well, a weekly check of ammonia levels could help you catch problems before they become disasters.

Add water if necessary to bring the water level of your tank back up

Unless it is extremely warm, I've found that topping off your fish tanks once a week is generally sufficient to keep your water levels from dropping too low. If you are adding more than about 10 percent of the total water volume, be sure the water is dechlorinated, and that you don't change the pH or the temperature of the water dramatically.

Check for insects

You can do this while your fish tanks are filling up, or any time you harvest a plant. Insects will generally live on the underside of leaves or in the stem junctions wherever tender new growth is emerging. Early identification is the key to successful pest control.

Monthly

Clean out the pump and the pipes

Yes, this is a dirty, nasty job, but it needs to get done. On a monthly basis, you could unplug your pump, take apart your pipes, and run high-pressure water (preferably dechlorinated, so you can preserve the bacteria in your pipes) from a hose nozzle through each plumbing component to prevent clogged arteries from solids buildup. The good news is that fish solid waste has no aroma!

"When to Hit the Ammonia Panic Button"

At some point in every system, there is a bit of a wobble in the nitrification cycle — perhaps you have been overfeeding, have not been checking on the system regularly or whatever. What I have noticed as a trend is that people appear to be very jumpy about ammonia and tend to go into panic mode at concentrations that I typically ignore in my system.

Your choice of fish is important here, I know. Mozambique tilapia are tough bastards alright — reason enough for me to recommend them to anyone that is in an area where they can be legally and responsibly owned. In a past "scare," I had nitrites touching 7 ppm and ammonia over 5 ppm. I did not lose a fish and I did not make even a partial water change. I was worried, but I had a couple of data sets at my disposal which I used as a "Plan B" generator in the situation. Water was a premium to me, not only because of all of the nutrients already in the AP system, but because of

the drought I'm working in. Water loss was therefore not a great idea for me. What I did have was the tolerances for the species (water quality) from a reliable source — aquaculture publications. I also have a decent knowledge of how temperature and pH work together to influence the toxicity of ammonia and nitrite.

Thus plan B was keep the temp down, the DO levels high, pin the pH to 6.5 and slow the feeding rate right down. It took a full 20 days before the readings were such that I thought the system was good to go again, but the moral of the story for me was that there were more than one way to get a system out of a tail spin, and that it is not always necessary to go for the parachute.

— Kobus Jooste,
Aquaponic System Development, South Africa

Agitate the solid waste accumulation in the fish tank

If you don't have a round fish tank with a conical bottom or excellent water circulation within the tank, chances are solid waste is accumulating in the bottom of your tank. It's a good idea to get in there monthly and stir it up while the pump is on so the pump can move it up into the grow beds.

Check nitrate levels

Nitrates are generally a good thing, right? Well, yes and no. Nitrates above 150 ppm could indicate that you do not have enough plants to take up the nitrogen that is being released by the nitrifying bacteria. If you see your nitrate levels rising, it's time to get more plants into your system, or harvest some fish. Either is generally good news in my world! If you still see them rising you might want to consider adding additional grow beds.

Aquaponic System Maintenance Rules of Thumb

- Ammonia, nitrites, nitrates (after cycling).
- Ammonia and nitrite levels should be less than or equal to 0.5 ppm.
- If you see ammonia levels rise suddenly, you may have a dead fish in your tank.
- If you see nitrite levels rise, you may have damaged the bacteria environment in your system.
- If either of the above circumstances occurs, stop feeding your fish until the levels stabilize, and, in extreme cases, do a one-third water exchange to dilute the existing solution.
- Nitrates can rise as high as 150 ppm without causing a problem, but much above that, you should consider harvesting some fish and/or adding additional plants or another grow bed to your system.
- Keep pH around 6.8–7.0. If it falls to 6.4, buffer it up.

In Conclusion

You know you are addicted to aquaponic gardening when ...

Because you purchased this book you probably already had some level of interest in growing fish and plants together before you even started reading. My sincere hope is that now that interest has blossomed into a passion and you are confidently working on your new aquaponics system.

Aquaponic gardening really starts when your system is cycled and the nitrifying bacteria are converting the fish waste into plant food. But aquaponic gardeners are often more than merely people who set up aquaponic systems. Many exude intensity and an evangelistic spirit about their gardens. They sense that they are on the verge of something new, big and world-changing — and truly addictive.

So are you wondering how you will know when you have become an aquaponic gardener? Here is a checklist that may help you pinpoint the moment when you cross the line from hobby to passion to fanatic.

You know you are addicted to aquaponic gardening when...

- Your fish recognize you.
- You buy a special home for the BSF maggots you are growing to feed your fish.
- You are known throughout the neighborhood as the house that grows the fish.
- You know the temperature of your fish tank every single day.

203

- You talk about bacteria aging like others talk about fine wines.
- You banish your car to the driveway so you can have more room to grow in the garage.
- You stress out over a tenth of a point move in pH or ammonia.
- You buy a bigger truck so that you can haul more gravel.
- Your non-gardening spouse becomes conversant in aquaponic terms.
- You look forward to a Saturday afternoon spent at the dump looking for old bathtubs and barrels.
- You plan vacation trips around visits to other aquaponic gardeners.
- You travel 150 miles to pick up 20 tilapia.
- On your monthly budget, plants and fish are more important than groceries.
- You could swear your fish are on the verge of speaking to you.
- The terms "bacteria", "poop" and "waste" take on entirely new meanings.
- You would rather go shopping at an aquarium store than just about any other store.
- You take every single person who enters your house on a "garden tour."
- You look at cattle trough and think "grow bed."
- You ask for tools for Christmas, Mother's/Father's day, your birthday and any other occasion you can think of.
- Invasive pond life like algae and duckweed are now treasured fish food.
- You know how many bags of Hydroton® your car will hold.
- Your preferred reading matter is forum threads.

And last but not least:

You know that the four seasons are:

- Planning your System
- Cycling your System
- Maturing your System ~and~
- Expanding your System

When I posted this on the Aquaponic Gardening Community I got some wonderful additions to the list...

- "The local hardware stores think you are a plumber because of all of the PVC pipe and other supplies you constantly buy." — Perry Adkisson, Bakersfield, California

- "Your spouse suggests putting a bed in the greenhouse. But you don't realize they are talking about a bed for you to sleep in." — Wayne, Bladenboro, NC
- "Several people say you should convert your 10,000-gallon swimming pool to a fish tank and you actually consider it." — Raychel A Watkins, Waianae, Hawaii
- "You find yourself looking around your property and thinking, 'I could squeeze a small system onto the patio. I wouldn't have anywhere for company to sit, but that's okay..,'" —Andrea, Yucaipa, California
- "You get up in the middle of the night and go outside in your PJs with a flashlight ... just to make sure everything is OK." — David Hart, Saint Cloud, Florida
- "You keep a funnel and empty bottles handy to save 'hummonia' for system cycling." — TCLynx, Tangerine, Florida
- "You start working on your systems at sunrise, skip meals, and at sunset wonder where the day went." — Chris Smith, Kailua Kona, Hawaii
- "When walking through the city you imagine the trajectory of the sun to find THE perfect sunny side of any building." — Rainey Sastrowidjojo, Paramaribo, Suriname
- "At the sight of any plastic container you wonder, 'is that food-grade plastic?'" — Jim Reed, Bridgewater, Massachussets
- "When it is time to upgrade your cell phone, you go for a waterproof model because you have your hands in water or duckweed beds so often you lose calls." — Kobus Jooste, Uitenhage, Eastern Cape, South Africa

If you are nodding your head, smiling and saying, "Yep, that's me!" then welcome to the community! We are thrilled that you have decided to join us on this adventure of growing your own food using this extraordinary method called aquaponic gardening.

Appendices

Appendix 1: Troubleshooting

"Good Food, Clean Hands, Happy Heart. Aquaponics."

— Richard Wyman, Aurora, Colorado

An aquaponic system is a living, breathing ecosystem that may occasionally become "sick" and perhaps even out of balance. Solutions to most of the problems that I have ever encountered in my aquaponic systems are outlined below. If you run into a problem that isn't specified here, my best advice is to join an aquaponics community site (see a list in Resources) and search their discussion threads for symptoms similar to yours. If you cannot find what you are looking for, post a new discussion thread. Some other aquaponic gardener is bound to have experienced the same issue, and hopefully solved it.

Cycling

I've been adding ammonia for several weeks now but I still haven't seen any trace of nitrites.

Unless there are ideal conditions present for bacteria to colonize, this may not actually be a problem. That said, here are some tips for speeding up cycling:

- Add nitrifying bacteria to your system — The best options here are either the filter sponge from an already cycled tank or media from a friend's aquaponic system. Don't forget that the nitrifying bacteria require a surface to establish. Just using water from another fish tank won't necessarily do the job.

- Make sure that you keep your ammonia levels at about 5 ppm. Too much ammonia can inhibit cycling.
- Increase the temperature of the area you want the bacteria to colonize by increasing the temperature of your water. Optimal temperature for bacterial reproduction is 77–86 °F (25–30 °C).
- Keep pH between 6.0 and 8.0 to facilitate growth of your bacteria colony.
- Provide plenty of aeration.

Plumbing

My grow beds are dry and it seems like my pump isn't working.

Are you using a timer? If so, there are two things that could be causing your grow beds to dry out. Either you have a malfunctioning pump or a malfunctioning timer. Flip your timer override switch to the "on" position. If your pump comes right on then it is probably your timer that is faulty and needs replacing.

Are you using a siphon? Then timers are not in the equation for you and you most likely have a malfunctioning pump.

My siphon isn't firing.

This means that there isn't enough water volume flowing through the pump. If your siphon was working well, then stopped working, this is probably because there has been a buildup of fish solids in your plumbing that is restricting the flow of water through the pipes just like a restricted blood flow through a clogged artery. To solve this problem you should take apart your plumbing and clean it out with a high-pressure stream of water.

My siphon won't stop.

This means that there is too much water going through your siphon. Did you just clean out your pipes? That could increase the water flow. The solution is to use the valve on the pipe delivering water to your fish tank to decrease the flow of water. If you haven't installed a valve on that pipe, now might be a good time to do so.

Fish

My fish aren't eating.

This is usually the first sign that something is wrong with your aquaponics system. If your fish aren't eating normally, you should take it seriously and try to figure out what is going on right away. The following are some of the issues that could cause fish to stop eating.

- Lack of oxygen — while they might not be so deprived that they are gasping at the surface, low levels of oxygen will cause them to not feel quite right.
- Lack of filtration resulting in increased ammonia levels (for example, if the water in your fish tank isn't being filtered once an hour).
- Water temperature is dangerously low or high for the fish type you are growing.
- Stress. Some possible causes of stress in fish are being handled, too much noise, too many people or an aggressive fish in the tank that is trying to defend its territory.
- pH being outside of optimal range or changing suddenly.

My fish are dying.

This is most likely being caused by any of the issues described above that have been causing your fish not to eat. Any item on that list can cause long-term physiological damage to your fish or suppress their immune systems. Also, I would add disease to this list if your fish are dying. Inspect any dead fish carefully for potential signs of disease and either take them to your local aquarium expert for advice or do some internet research.

My fish are behaving aggressively.

If your otherwise docile fish start behaving aggressively toward each other, it is probably because there has been a disturbance in the social order of the school. Perhaps a new fish has been introduced, or they have been moved to a new tank? Perhaps a female is brooding and the males are trying to establish nests. Whatever the reason, you need to deter this behavior as quickly as possible by giving the victim(s) a place to hide. I've found that garden pots turned on their sides, or 12-inch to 18-inch (30–45-mm) pieces of six-inch (15-cm) diameter PVC pipe, work well for this purpose.

My fish are hanging around on the tank surface and gulping.

You have an oxygen problem in your tank. Are your oxygenating devices working? Is the siphon triggering? You need to solve this problem quickly because even a brief shortage of oxygen can cause long term gill damage to your fish.

My fish have odd spots or growths on them.

If you see any odd discharge or growth on your fish, you probably have a fish disease. Isolate or destroy that fish immediately and figure out what you have. You may need to treat, or replace, the entire tank's population.

Plants

My plants are wilting.

There are a number of possible causes for this problem:

- Over-watering — More accurately defined as lack of oxygen. This could happen if the water isn't draining properly from your grow bed if your siphon fails to fire or your timer doesn't advance to the "off" position. The plumbing where the water exits the grow beds might also need a thorough cleaning so the water drains more quickly from the grow bed.
- Under-watering — This could happen if you have a pump or a timer problem. The media will stay moist for quite a long time, however, so if this is causing your plants to wilt you actually have a much bigger problem with the health of your fish from lack of filtration.
- Insects — If harmful insects are allowed to take over a plant they could cause it to wilt and die. Inspect under the leaves of the plants and in the new growth junctures for insects. Also check around the base of the plant for fungal gnat larvae, which sometimes take up residence in soil-less media and eat the roots of plants.
- Disease — Plant diseases are rare in aquaponics, but they do exist. If the plant is being watered properly and shows no sign of insect infestation, this is the only remaining option. Pull the plant immediately and destroy it.

My plants are not a deep green color. The leaves have a yellow cast to them.

This is probably being caused by a nutrient deficiency. This is actually fairly common during cycling and the first three to four months of an aquaponic grow bed's life. The bacteria haven't developed enough yet to convert enough fish waste into plant nutrients. After this initial start-up period has passed, you should no longer see nutrient deficiencies. As long as you keep the pH close to 6.8, use a high-quality fish feed and have stocked to proper levels, this problem should right itself. If you continue to see it, however, you might have an iron deficiency. Adding a small amount of chelated iron powder to your system should solve this in a week or so.

The leaves are falling off my plants.

This could be caused by any of the factors that cause wilting or nutrient deficiencies above. Dropping leaves is a plant's way of announcing that it is highly stressed and is preparing for the end of its life.

My plants have a powder all over them.

Your plants have probably been attacked by a fungus called powdery mildew. I find it is especially common on braising greens such as Swiss chard and cucumber plants. There are organic sprays available, and the neem oil you might be using for insect prevention will help prevent powdery mildew as well. Also, potassium bicarbonate makes an excellent antifungal spray as well as an excellent pH buffer. It can be found in beer brewer supply shops.

Water chemistry (post cycling)

My pH has dropped below 6.4.

This is perfectly normal in a healthy aquaponics system. The nitrification process has a downward pulling effect on pH. At this point you should add an alkaline buffer, such as hydrated lime, to bring your pH back up to 6.8–7.0; see the Water chapter for more details.

My pH has risen suddenly.

There are a few reasons why this may happen:

- An anaerobic, or "dead," spot in your grow beds caused by solid waste can constrain or stop good water and oxygen flow in that area. You will need to find out where this is happening (hint: stir your media and use your nose to find an area of rotting smell) and clean it out.
- Changes in water supply — Municipal water districts often draw water from different sources at different times of the year. Perhaps your municipality has recently switched to a more alkaline, harder source of water.
- Slow release agent in your media — If you have added seashells or eggshells or some other natural pH buffering source to your media, you may find that it suddenly starts decomposing and affecting your pH.

My ammonia levels are suddenly very high.

Assuming you were fully cycled and your ammonia levels were at or near zero, a sudden spike in ammonia could be caused by the following:

- Dead fish — Decomposing flesh gives off a great deal of ammonia.
- Overfeeding — Decomposing fish food can also be a source of ammonia. Make sure you only feed as much as your fish will eat in five minutes, and remove any uneaten food from the tank.
- Lack of filtration — Is the water flooding all the way through your grow bed when the pump goes on? Is the entire volume of water being filtered once an hour?

My nitrate levels are above 100.

High nitrate levels are a sign that you can support more grow beds. This is good news!

Appendix 2:
Aquaponic Gardening Rules of Thumb

What follows is a list of all the basic "rules of thumb" from the summaries at the end of each chapter in the book. This is intended to be a place for you to easily refer if you have a quick question about your aquaponics setup.

I had the distinct honor of collaborating with Dr. Wilson Lennard from Australia on these guidelines for beginning aquaponic gardeners. Dr. Lennard has earned one of the few PhDs in aquaponics in the world, and he currently runs a consulting practice called Aquaponic Solutions. These guidelines have also been reviewed and endorsed by Murray Hallam of Practical Aquaponics, and the *Aquaponics Made Easy* and *Aquaponics Secrets* videos, and been vetted by the Aquaponic Gardening Community.

Nothing we say below is set in stone and there are exceptions to almost every one of the listed rules of thumb given certain conditions. However, they do offer a set of generally accepted principles that, if adhered to, will put you on the path toward successful aquaponic gardening.

Aquaponic System Design Rules of Thumb

I recommend a media bed for new, hobby growers. Why not NFT or deepwater culture (aka raft or DWC)?

A media bed performs three filtering functions:
- mechanical (solids removal)
- mineralization (solids breakdown and return to the water)

- biofiltration
- Because the media bed also acts as the place for plant growth, it basically does everything all in one component — making it all simple.
- Media also provides better plant support and is more closely related to traditional soil gardening because there is a medium to plant into.
- The cost of building the system is lower because there are fewer components.
- It is easier to understand and learn.
- Basic Flood and Drain is the simplest system to design and is appropriate for a 1:1 grow bed to fish tank volume.
- CHOP or 2-Pump systems have sump tanks and enable a 2:1 and even up to a 3:1 grow bed to fish tank ratio. More grow beds filtering the water is generally better for the health of the fish.

Aquaponic Grow Beds and Fish Tanks Rules of Thumb

Steps for planning your system

Determine the total grow bed area in square feet

- From grow bed area, determine the fish weight required (in pounds or kg) using the ratio rule 1 lb (500 g) of fish for every 1 sq ft (0.1 m²) of grow bed surface area, assuming the beds are at least 12 inches (30 cm) deep.
- Determine fish tank volume from the stocking density rule above (1 lb fish per 5–10 gallons of fish tank volume or 1 kg of mature fish per 40–80 liters). When your fish are young and small, reduce the number of plants in proportion to the size of the fish and their corresponding feed rate / waste production.

Grow bed and fish tank

- Start with a 1:1 ratio of grow bed volume to fish tank volume. You can increase that up to 2:1 once your system starts to mature (4–6 months) if you want to.
  - Must be strong enough to withstand the lateral and downward forces of the media, water, and plant roots.
  - Must be made of food-safe materials and should not alter the pH of your system.

Grow bed

- The industry standard is to be at least 12 inches (30 cm) deep to allow for growing the widest variety of plants and to provide complete filtration.

- Be sure to create or purchase a media guard to facilitate easy cleaning of your plumbing fittings.

Fish tank

- If you have flexibility here, 250-gallon (1000-liter) or larger seems to create the most stable aquaponics system. Larger volumes are better for beginners because they allow more room for error; things happen more slowly at larger volumes.
- You need at least a 50-gallon (200-liter) volume to raise a fish to 12 inches (30 cm) ("plate size").

Aquaponic Plumbing Rules of Thumb

- You should flood and then drain your media-based grow beds. The draining action pulls oxygen through the grow beds. The least complicated way to achieve a reliable flood and drain system is to use a timer. While more complex, siphons are also excellent options for aquaponics.
- If you are operating your system with a timer you should run it for 15 minutes on and 45 minutes off.
- You should flow the entire volume of your fish tank through your grow beds every hour, if possible. Therefore, if you are running your pump for 15 minutes every hour, and you have a 100-gallon (375-liter) tank, you need at least a 400-gallon per hour (gph) (1,500-liter per hour) pump. Also consider the "lift" or how far against gravity you need to move that water and use the sliding scale on the pump packaging to see how much more power you need beyond the 400 gph.

Aquaponic Media Rules of Thumb

Any media you select

- Must be inert — i.e., won't alter the pH of the system
- Must not decompose
- Must be the proper size (½"– ¾" / 12–18 mm aggregate is optimal)
- The most widely used media types are LECA (lightweight expanded clay aggregate, aka Hydroton®), lava rock, expanded shale and gravel.
- If you choose gravel, understand its source and avoid limestone and marble as they could affect your pH.

Aquaponic Water Rules of Thumb

Purity

- Be sure to "off-gas" chlorine from your water before adding it to your system or use a chlorine filter.

Temperature

- If possible, select fish that will thrive at the water temperature your system will naturally gravitate toward.
- It is easier to heat water than it is to cool it.
- Attract heat by using a black tank or making it black.
- Retain heat through insulating techniques.

Oxygen

- Dissolved oxygen levels for fish must be above 3 ppm, and preferably above 6 ppm.
- You cannot have too much oxygen in an aquaponics system.

pH

- Target a pH of 6.8–7.0 in your aquaponic system. This is a compromise between the optimal ranges of the fish, the plants and the bacteria.
- Test pH at least weekly, and as frequently as 3–4 times per week, using your API Freshwater Master Test Kit.
- During cycling, pH will tend to rise.
- After cycling, your system's pH will probably drop on a regular basis and need to be adjusted up. If you need to lower pH, it is generally because of the water source (such as hard groundwater) or because you have a base buffering agent in your system (eggshells, oyster shell, shell grit, incorrect media).

Best methods for raising pH if it drops below 6.6:

- Calcium hydroxide — "hydrated lime" or "builder's lime."
- Potassium carbonate (or bicarbonate) or potassium hydroxide ("pearlash" or "potash").
- If possible, alternate between these two each time your system needs the pH raised. These also add calcium and potassium, which your plants will appreciate.
- While they work, be cautious about using natural calcium carbonate products (eggshells, snail shells, seashells). They don't do any harm, but they

take a long time to dissolve and affect the pH. So you add it, check pH two hours later and nothing has changed, so you add more. Then the pH suddenly spikes because you have added so much.

Best methods for lowering pH, in order of preference, if it goes above 7.6:

- pH Down for Hydroponics (be careful of using the aquarium version as this has sodium that is unhealthy for plants).
- Other hydroponic acids like nitric or phosphoric as the plants can use the nitrate or phosphate produced.
- Other acids such as vinegar (weak), hydrochloric (strong) and sulphuric (strong) — last resort, as directly adding these acids to your system could be stressful for your fish.
- Avoid adding anything to your system containing sodium as it will build up over time and is harmful to plants.
- Do not use citric acid as this is anti-bacterial and will kill the bacteria in your biofilter.

Aquaponic Fish Rules of Thumb

Stocking density

- 1 pound of fish per 5–7 gallons of tank water
- Fish selection should take into account the following:
  - Edible vs. ornamental
  - Water temperature
  - Carnivore vs. omnivore vs. herbivore

Oxygen needs

- When introducing new fish into your system:
  - Be sure your system is fully cycled
  - Match pH
  - Match temperature

Feeding rate

- Feed your fish as much as they will eat in five minutes, one to three times per day. An adult fish will eat approximately one percent of its body weight per day. Fish fry (babies) will eat as much as seven percent. Be careful not to overfeed your fish.

- If your fish aren't eating they are probably stressed, outside of their optimal temperature range or don't have enough oxygen.

Aquaponic Plants Rules of Thumb

- Avoid plants that prefer an acidic or basic soil environment. Otherwise just about any plant can be grown in an aquaponics system.
- Plants can be started for aquaponics the same way they would for a soil garden — by seed, cuttings or transplant.
- If your plants are looking unhealthy after the first few months it is probably for one of two reasons:
  - Nutrient imbalance caused by out-of-range pH — Maintain pH between 6.8 and 7.0 for optimal nutrient uptake by your plants.
  - Insect pressure.

Aquaponic Worms Rule of Thumb

- Add a handful of composting red worms to each grow bed once your system is fully cycled and fish have been added.

Aquaponic Cycling Rules of Thumb

Photocopy the data collection worksheet in the appendices to manage the cycling process. It is also available online at AquaponicGardening.com.

Cycling with fish

- Add only half as many fish as you would to be fully stocked.
- Test daily for elevated ammonia and nitrite levels. If either gets too high do a partial water exchange.
- Feed once per day or less to control ammonia levels.

Fishless cycling

- Sources of ammonia:
  - Synthetic — pure ammonia and ammonium chloride.
  - Organic — urine and animal flesh.

The process

- Add the ammonia to the tank a little at a time until you obtain a reading from your ammonia kit of 2–4 ppm.
- Record the amount of ammonia that this took, and then add that amount

daily until the nitrite appears (at least 0.5 ppm). If ammonia levels approaches 8 ppm, stop adding ammonia until the levels drop back down to 2–4 ppm.

- Once nitrites appear, cut back the daily dose of ammonia to half the original volume. If nitrite levels exceed 5 ppm, stop adding ammonia until they drop to 2.
- Once nitrates appear (5–10 ppm), and both the ammonia and the nitrites have dropped to zero, you can add your fish.
- pH should be between 6.8 and 7.0
- Plant your system as soon as you begin cycling. Adding liquid seaweed will help your plants quickly acclimate to their new environment.
- Adding bacteria will dramatically speed up the cycling process. Keeping the temperature of your water above 70 °F (21 °C) will help as well.

The Murray Hallam cycling technique

- Add liquid seaweed to the system.
- Add plants.
- Wait for two weeks.
- Then add fish.

Aquaponic System Maintenance Rules of Thumb

- Ammonia, nitrites, nitrates (after cycling).
- Ammonia and nitrite levels should be less than or equal to 0.5 ppm.
- If you see ammonia levels rise suddenly, you may have a dead fish in your tank.
- If you see nitrite levels rise, you may have damaged the bacteria environment in your system.
- If either of the above circumstances occurs, stop feeding your fish until the levels stabilize, and, in extreme cases, do a one-third water exchange to dilute the existing solution.
- Nitrates can rise as high as 150 ppm without causing a problem, but much above that, you should consider harvesting some fish and/or adding additional plants or another grow bed to your system.
- Keep pH around 6.8–7.0. If it falls to 6.4, buffer it up.

Appendix 3:
The top 10 dumbest mistakes
I've made in aquaponics

"Experience is the name everyone gives to their mistakes."
— Oscar Wilde

It's good to be able to laugh at yourself, right? Especially if you are laughing about dumb mistakes that you've made that you can tell others about — hopefully so that they will avoid making the same ones. It is with this spirit of philanthropy that I pass on to you, dear reader, the countdown of the top 10 dumbest mistakes I've made in aquaponics … so far…

10 — Get attached to my tilapia

I admit it. I'm a wimp. I love to eat fish, but I just can't seem to bridge the divide between that fish swimming in my tank and the one on my plate. I intellectually love the notion of raising my own protein, but I can't deal with the reality of executing these creatures that I've fed and fretted over all this time. I think I might need to bribe a friend to come do the deed someday on the promise of the freshest imaginable fish dinner.

#9 — Not jumping on bug problems in the greenhouse fast enough

I saw a few aphids, and then saw a few more. By the time I ordered predator insects, the aphids were everywhere and it took several batches of ladybugs at much expense that could have been avoided if I had just taken those first few aphids seriously.

#8 — Not testing ammonia once a week

Picture this scene. I had some students from an aquaponics class I taught in Boulder over to my greenhouse for some hands-on practical experience. I showed the class how to test pH, ammonia, nitrites and nitrates using a test kit. As I shook the vial containing the ammonia test I commented that there probably wouldn't be anything to see because this system had been running long enough to have a very well-functioning biofilter. Then the vial started turning green, and it just kept getting greener. After the required five minutes for a valid test had passed, it was clear that I had very high levels of ammonia in that tank. The cause was a dead fish back in the corner that I couldn't see (see mistake #4 below). Even with a partial water exchange, I lost several fish over the next few days. If I had been testing ammonia regularly I would have caught this problem early on.

#7 — Let the tank water get too hot

Tropical fish such as tilapia can handle very warm water provided there is enough oxygen. That is the rub. As the temperature of the water increases, its ability to hold oxygen plummets. I decided to put one of our new AquaBundance systems out on the corner of our deck last summer to see what would happen. It functioned perfectly, but when that high-altitude Colorado sun started beating down on it in the dog days of summer, the water temperature rose to 95 °F (35 °C). I realized too late that I needed to add extra oxygen to the water and I lost several fish over the next few days due to gill damage from lack of oxygen.

#6 — Too many fish in the system

I'm doing this right now. I have 70 4–6″ (10–15-cm) tilapia in about 100–120 gallons (380–450 l) of water. It looks like a super-highway during rush hour down there. My excuse for this is mostly related to mistake #4 below, but also to my general reluctance to kill fish to eat them. The theory was that by now I would have enjoyed the fully mature tilapia in the neighboring tank for dinner, thus making room for these smaller guys to spread out. Problem is, I can't reach them.

#5 — Used cheap "feeder" goldfish

We had dreams of growing beautiful koi in our newest tank, but thought we should start with goldfish until the system was cycled and we were sure there weren't any problems. To my delight I could get feeder goldfish for about 18 cents at the mega-chain pet store in town. I got about twenty of them and they died off, slowly but surely, one by one, over a few weeks. Thinking it was probably just cycling stress, I got another forty. Same thing happened. I found out that they probably came in with a disease (why keep them healthy when they are probably destined to be quickly eaten anyway?) and that my system needed to be sterilized and recycled. Lesson learned.

#4 — Set up my grow beds over my fish tanks so I can't access the fish tanks.

When we set up our greenhouse, the most space-efficient place to set up the grow beds was over the fish tanks. This is where we put the beds when it was a hydroponic system. With the grow beds over the fish tanks, the only remaining place for the plumbing was between the beds, basically taking up the remaining accessible room in the tank. Didn't account for the fact that tilapia are speedy little buggers and not very cooperative about being netted. Plus, I can't observe what is happening back in the far corners of the tank. Could be a crazy fish party back there … who knows?

#3 — Electrocuted the fish

Tank plumbing in greenhouse.

It's true. One summer evening while having dinner at a restaurant with friends I got a call from my then 14-year-old son saying, "Ah, mom, how come I got shocked when I stuck my hand in the fish tank?" That put a damper on a relaxing dinner, I can assure you. Ends up there was a plug on the ground near the tank that got wet. Won't do that again. The fish? Well, they stopped swimming and we thought we had lost them but once the shock wore off they snapped out of it and recovered just fine. Tilapia are amazingly hardy.

The Aquaponic Source, Inc.

#2 — Swung the pH in the fish tank from 8.0 to 6.8 in five minutes

I was actually at an aquaponics workshop over Memorial Day weekend having just set up my first aquaponics system about three weeks earlier. This was a joint project with my son so he and I talked frequently while I was gone and I had him doing pH and other testing on the system daily (side note — it's great to have excuses to communicate with your teenager, no matter what the topic!). He told me the first morning of the workshop that the pH had climbed to 8.0. I knew that was a big problem so told him how to take the pH back down to 6.8. I then hung up the phone and went into the class. Within an hour we were instructed to ALWAYS take pH down very gradually, i.e., over a period of days. Oh no! At the break I got my son back on the phone but it was too late. He had dropped the pH in our system from 8.0 to 6.8 in a matter of minutes. I think we lost one or two tilapia in that episode. Like I said, they are incredibly hardy fish.

And the #1 dumbest thing I've done — Run tap water into my fish tank all day because I forgot to turn off the hose.

I was at work when I got the call at about four in the afternoon: "Ah, mom, I just turned off the hose that was filling up the fish tank. Water was spilling out all over and I think it has been running all day." Oh no! I had meant to top off the fish tank in the morning with just a little water, had been distracted, and had left the hose running chlorinated water through the system ALL DAY! Lost about 14 fish and all of my beautiful bacteria. Had to cycle the entire system again. Devastating. Now when I use the hose (i.e., a chlorinated water source) I stick something large in my pocket so the discomfort reminds me of the hose... and the "discomfort" that my fish will feel if I leave it running.

Appendix 4: What to consider before plunging into commercial aquaponics

"Farming is going to be one of the great industries of the next 20 years or so. I'm wildly bullish on agriculture. Farmers will be the ones driving the Lamborghinis in the future."

— Jim Rogers, Rogers Holdings Chairman, *Bloomberg* interview

I have people contact me almost daily who have just come across this amazing new growing method called aquaponics. Much of the time they are considering becoming a commercial aquaponics farmer. This aspiration is generally connected to a career change and a new direction in life. Commercial aquaponics can be a very seductive, romantic "siren song" if your days are spent in uninspiring pursuits. I know. I was there.

When I discovered aquaponics, the first thought I had was to start a commercial operation. This led me to months of researching, analyzing and interviewing before I finally concluded that it just wasn't the right direction for me. However, I learned a lot along the way and I'd like to pass some of those lessons on to you just in case you are considering the same idea. Perhaps it truly is the right direction for you.

1. Know your consumer!

Your market will be local, and you will be competing with large growers that can cheaply produce and probably bring in product from all over the world. With aquaponics you probably won't be able to compete on price but you will be able to offer organic, sustainable and "local" produce. So, the key question you must answer is, "Will your local consumers pay a premium for the distinct value you will be offering?" If the simple answer is "yes," then you will have identified a market. But the next question is, "What is the size of that market and is it big enough to sustain your business model?" If your analysis tells you the market is big enough, ask yourself what else might impede the market accepting your product? Here is an example of what I mean. Organic, local produce is highly sought after in my hometown of Boulder, but along with that enthusiasm comes a romantic attachment to traditional dirt farming. People here often automatically think, "Oh, aquaponics is associated with hydroponics, which is very unnatural and is connected to pot-growing. Yuck." Overcoming this perspective would require some serious marketing. Bottom line — do your homework, be honest with yourself about what you discover, and plan your business to best meet the needs of your market.

2. The further away from the end consumer you get, the more convenient it is for you, but the more people will be taking a slice of your profits.

We discovered you could market local organic produce directly to the end consumer through farm stands, farmers markets and CSAs (Community Supported Agriculture; see point 5 below). But selling direct limits the amount you can sell and therefore the sales volume you can generate. However, if you want to use various distribution channels, you need to sell large volumes at much lower margins in order to become profitable. This suggests that you need to decide early what kind of business you wish to build — small and local with higher margins but lower sales, or bigger with a wider reach but lower margins.

3. Talk to lots of people in the industry, especially potential buyers

This was incredibly useful. I talked to several restaurant owners and chefs, and learned that they could be a very viable market for picked and delivered-that-day produce, especially salad greens. Chefs were also very interested in being able to specify what was grown for them. This suggests there might be an interesting market for custom salad blends and unusual herbs. I also talked with grocery store managers in town and found that one family-owned store would buy everything I grew (price would probably be an issue, though) and another, Whole Foods, loved working with local producers (it is a big focus for them), loved the idea of promoting aquaponics (we even discussed having a web-cam link to the greenhouse so customers could see the activity there, again MARKETING is critical) and they did not have a local supplier of watercress so were very interested in that. All great learning.

4. Consider focusing on off-season produce

We have a very active farmers' market (biggest in Colorado and considered one of the top 10 in the country) but given our climate, everything is very seasonal, and fresh produce goes away for about half the year. There is a big market for off-season produce. Lettuce in the middle of the summer. Tomatoes in April. Everything from October to April. If you can figure out how to keep heating costs down, this could be the entire focus of a business plan right there.

5. Consider a CSA model

This is basically an extension of everything discussed above. After all was said and done, we decided that the best way to approach our market was through a year-round CSA (Community Supported Agriculture) model where we would sell shares of our production to "members." The members would be invited to participate in the decision-making about what we would grow, and would get a percent allocation of what came out of the greenhouse each week (including fish) — and if there was a catastrophe they would share in that as well. The shares would be in three-month blocks, and the members could pick their own produce from a list we would provide, if they wanted to (we thought families with kids would love this), or we would pre-pick for them and a box would be ready for pickup on certain days. Any "excess" produce could be sold at the greenhouse, sold to restaurants, donated to the food bank or fed to the fish.

6. Consider the flexibility of your system

Given the CSA model we were leaning toward, being able to grow a wide variety of crops was extremely important to us. I was designing our greenhouse with a combination of media beds and vertical towers for fast-growing plants. I moved away from using raft systems because of the limited set of plants you can grow, and because they are more expensive to install and take more time to maintain. I believe it is important to be nimble and be able to react to your market no matter which model you start with.

7. Fish probably won't be a profit center

Here in Colorado we have a federal prison that produces most of the "local" tilapia sold in grocery stores. It's pretty tough to compete with prison labor. The price that tilapia fillets earn barely covers the cost of feed. We found that some grocery stores were interested in perch and barramundi, but these fish are more difficult to raise (maybe not the perch) and grow relatively slowly. The bottom line with fish is that we were hoping to break even on the cost of feeding them and heating their water. In exchange, they would provide all of our plant food at no additional charge! Plus, they would add a very potent point of differentiation to our organic produce operation. Look into your local market prices for the fish you want to grow as part of your analysis.

8. You may not be able to "process" your fish

In order to fillet a fish in Colorado, and I believe most of the rest of the country, you need to pass through some serious regulatory hoops and have a specially designed and inspected facility. You can, however sell them live, or "mostly live," meaning whole and on ice. When we looked into this, it meant that we could only sell to Asian markets and/or directly to consumers and restaurants. Again, restaurants were the most excited but only if they could have the fish they wanted.

Tilapia wasn't all that enticing to them. Again, figure out what works in your market and what regulations you are going to have to face.

9. Start with a small setup and learn first

I can't say this strongly enough. The riskiest thing you can do is go to a weekend workshop and think that you are ready to manage a large aquaponics operation. While aquaponics is not too complex once you understand how it all works, you need to grow for a while before you become experienced at recognizing signs of trouble. I can now walk into my greenhouse and instantly know that something is wrong because the sounds and the smells aren't normal. I can look at the fish and observe their eating habits and tell how healthy they are. I can look at the plants and tell if I have a pH, nutrient or insect problem. When I first started I had none of this knowledge. I shudder to think what would have happened had I started a large growing operation before honing these skills.

10. Know thyself

I ultimately decided not to pursue a commercial aquaponics operation because I finally came to the personal realization that, although I have tremendous respect for farmers, I am not a farmer by nature. I don't do well with routine and solitude. I was intimidated by the notion that I had to be on call 24 hours a day until our business got big enough to hire employees, and even then would still be on call 24 hours a day, really. I'm a bit of a wimp when it comes to physical labor in a hot environment. Given my background, I'm vastly more comfortable with backyard systems and helping people successfully start aquaponics in their homes. I love to teach, write and promote — I'm a big picture person, not a detail person. But that is just me. Do you have what it takes to be an aquaponic farmer?

If after all this you still decide that commercial aquaponics is right for you, then my hat is off to you. The world needs more farmers, and I wholeheartedly support any and all efforts to establish aquaponics as a viable agricultural technique. My final bit of advice, however, is to invest in a reputable training course. How do you find one?

Doing a Google search and reading website content will only tell you that they are good at search engine marketing and designing websites. You should also join an online community site and ask questions of the members. Find out who is highly regarded and why. Aquaponics is still a very small world and with today's online transparency, separating the scams from the gems is not difficult.

Next, contact the training company, or companies, and ask for references. You will want to talk to former students who have taken the course and then started and run a profitable aquaponics farm for at least a year.

Finally, try not to weigh the price of the course too heavily in your decision. The low cost option might turn out to be an introduction intended to induce you to buy future equipment and consulting services rather than a complete course. Ask if they have successfully run a profitable, commercial aquaponics operation themselves or simply run a successful training and consulting company. You need hands-on education, not just theory.

Once you have addressed the issues above, created a business plan, taken a course or two and gotten some hands-on experience through a smaller aquaponics system, you will be on your way to becoming a successful aquaponic farmer. It won't be easy, but you can take pride in knowing that you are spending your days as an important part of the solution to global hunger, peak oil and climate change.

Appendix 5:
Aquaponics System Maintenance Checklist

(available as a downloadable PDF on AquaponicGardening.com/free-downloads)

Month of _____

Daily tasks: Feed fish, check fish tank temperature, check oumps and plumbing

	Monday	Tuesday	Wednesday	Thursday	Friday	Saturday	Sunday
Week 1							
Week 2							
Week 3							
Week 4							
Week 5							

Weekly tasks

	Week 1	Week 2	Week 3	Week 4	Week 5
Check pH					
Check ammonia					
Add water					
Check for insects					

Monthly tasks

Clean out pumps and plumbing	
Agitate solid waste accumulation	
Check nitrate levels	

Appendix 6:
Aquaponics System Data Tracking Sheet

(available as a downloadable PDF on AquaponicGardening.com/free-downloads)

Date	pH	Ammonia	Nitrite	Nitrate	Notes

Date	pH	Ammonia	Nitrite	Nitrate	Notes

Appendix 7: Recommended resources

Books

Hydroponic gardening

Hydroponic Food Production by Howard Resh. Commonly referred to as "The Bible" of hydroponics, this academic tome contains everything you need to know about how plants grow in a soil-less environment.

The Best of Growing Edge, Volumes 1–3. Growing Edge was a very special magazine. It was once described as "more than just another gardening magazine; it represents a different way of looking at the process of cultivating plants." Early issues included articles on topics like "Growing Bananas Indoors" and "The $50 Greenhouse." It was published from 1989 through 2009, and my son and I were on the cover of the May/June 2007 issue. These three article compilations from *Growing Edge* represent the some of the best thinking in alternative gardening over two decades.

How-To Hydroponics by Keith Roberto. This is the book I turn to for specifics about how to build hydroponic systems. With some adjustments for aquaponics (12-inch / 30-cm deep beds, for example) this is a terrific how-to guide for constructing the plant side of your system.

Aquaculture

Small Scale Aquaculture by Steven D. Van Gorder. The best book I've found for explaining how to set up and take care of the fish side of your aquaponics system.

Recirculating Aquaculture Systems by Michael B. Timmons, Ph.D. This definitive guide to all aspects of managing a recirculating aquaculture system was the primary resource for the Cornell University short course on aquaculture. The latest edition includes a chapter on aquaponics by Dr. James Rokocy of the University of the Virgin Islands.

Greenhouse design and gardening

The Food and Heat Producing Solar Greenhouse by Bill Yanda and Rick Fisher. This is a comprehensive greenhouse design book from 1979. There are over 60 pages of photographs of solar greenhouse examples. A classic for anyone planning to design and build the most efficient solar-heated greenhouse possible.

The Greenhouse Gardener's Companion by Shane Smith. This book is indispensable for any greenhouse gardener. It is where I turn for information on insect control, for example.

The Hydroponic Hot House: Low-Cost, High-Yield Greenhouse Gardening by James B. Dekorne. This book is a wonderful exploration of sustainable greenhouse growing without soil. It even includes a chapter called "Aquaculture" at the end. Since it was written in 1992, the term "aquaponics" hadn't been coined yet, and he never does connect his fish system to his plant system, but he was generally heading in the right direction.

Gardening in Your Greenhouse by Mark Freeman. Another practical guide to the unique aspects of greenhouse gardening.

Four-Season Harvest by Eliot Coleman. This book has a good greenhouse section and discusses what plants grow best under a variety of seasons and lighting conditions.

Plant propagation

The New Seed Starter's Handbook by Nancy Bubel. The most comprehensive book available on all aspects of seed starting.

Making More Plants: The Science, Art and Joy of Propagation by Ken Druse. This is a gorgeous book full of color photographs showing you every imaginable way to inexpensively multiply your plant stock from saving seeds to cuttings to grafting and divisions.

Creative Propagation by Peter Thompson. This is a 359-page, small print, intense guide to all aspects of propagation.

Breed Your Own Vegetable Varieties by Carol Deppe. For the serious plant nerd. Includes instructions on how to cross-seed varieties and create entirely new varieties of your favorite plants.

Urban homesteading

Complete Idiot's Guide to Urban Homesteading by Sundari Kraft. I worked with Sundari on the aquaponics chapter for this book, so I know the care she took writing it.

Animal, Vegetable, Miracle by Barbara Kingsolver. An entertaining romp through a year of a family trying to eat entirely locally. Extremely well written and engaging.

Farm City: The Education of an Urban Farmer by Novella Carpenter. A wonderful story of a city gal who decides to create a farm on her small urban lot in Oakland. The only thing missing was fish!

Global food production

The Vertical Farm: Feeding the World in the 21st Century by Dr. Dickson Despommier. An inspirational guide to one view of how we will produce food as we move more into our urban centers.

Just Food by James E. McWilliams. Contains "The Blue Revolution," an excellent chapter on the potential impact of aquaponics on aquaculture.

The End of Food by Paul Roberts. A detailed tome about what is wrong with our current global food supply system. No seed is left unturned.

Hot, Flat and Crowded by Thomas Friedman. A detailed tome about how the world is changing through globalization, climate change and increasing population.

Websites

Aquaponic Gardening — aquaponicgardening.com — the website for this book. Contains information on media based aquaponic gardening, resources, events, community and more.

The Aquaponic Source — theaquaponicsource.com/ — My website. Full of aquaponics information, resources and a full line of aquaponics products, supplies and accessories.

Travis Hughey of Faith and Sustainable Technologies, Barrel-ponics® manual — www.fastonline.org/

Aquaponics Monster Directory — www.backyardaquaponics.com/component/content/article/4/44-designs.html

SOS Planet Public Library—documents.ponics.org/category/aquaponics/
Sources for expanded shale — www.escsi.org/membermap.aspx

Forums and community sites

Aquaponic Gardening Community — aquaponicscommunity.com/ — The largest, most active online aquaponics community in North America.

Practical Aquaponics Forum — www.aquaponics.net.au/forum/ — Murray Hallam's aquaponics forum.

Backyard Aquaponics Forum — www.backyardaquaponics.com/forum/ — Joel Malcom's aquaponics forum.

Barrel-ponics® Yahoo Group — tech.groups.yahoo.com/group/ barrelponics/ — Travis Hughey's Barrrel-ponics® forum.

Blogs

Aquaponic Gardening Blog — theaquaponicsource.com/blog/ — This is my blog. I update it at least weekly with news and stories about aquaponics in North America.

Murray Hallam's Practical Aquaponics Blog — aquaponics.net.au/blog/ — Murray Hallam's blog. Filled with tips and stories about the aquaponics life in Australia.

EcoFilms — aquaponics.net.au/blog/ — Frank Gapinski of EcoFilms was bitten by the aquaponics bug when he first created Murray Hallam's *Aquaponics Made Easy* video and has been writing about his own experiences ever since.

Affnan's Aquaponics — affnan-aquaponics.blogspot.com/ — Affnan is known in the aquaponics world as the king of the bell siphon.

Other social media

Facebook — www.facebook.com/aquaponicgardening — the Facebook page for this book.
www.facebook.com/TheAquaponicSource
Twitter@aquapon — the Twitter account for Aquaponic Gardening.

Heirloom seed companies

Seed Savers Exchange
Seeds of Change
Southern Exposure Seed Exchange

References

AFP. "Lack of Crop Diversity Threatens Food Security: UN." *The Independent*, UK. October 27, 2010. www.independent.co.uk/environment/lack-of-crop-diversity-threatens-food-security-un-2117628.html (accessed January 2011).

Arancon, Norman. Clive A. Edwards, Richard Dick and Linda Dick. "Vermicompost Tea Production and Plant Growth Impacts." BioCycle, 2007: 51–52.

Astyk, Sharon. "Reconsidering Cities." sharonastyk.com/2010/01/12/reconsidering-cities/.

Backyard Aquaponics. "Fishing For Answers." September 2009: 12–15.

Batis, Jeff and Anwar Kaelin. "History of Aquaculture." www7.taosnet.com/platinum/data/medium/whatis/whatis.html (accessed December 2010).

Bernstein, Sylvia. "Beginning and an End: An Intervew With Dr. Jim Rakocy." *Backyard Aquaponics,* Januar, 2010, Q1:1, 34–35.

Boulder Daily Camera. "India's Farmers Say Climate Change Hurting Tea Growers." January 1, 2011.

Bradtke, Birgit. "How Does Neem Insecticide Work?" www.discoverneem.com/neem-oil-insecticide.html (accessed 2011).

Brown, Lester R. *Plan B: Rescuing a Planet Under Stress and a Civilization in Trouble.* Earth Policy Institute, 2003.

Cal Ag Ed. "Agriculture Core Curriculum: Cost Efficiency of Production." California Agricultural Education. May 2, 1990. www.calaged.org/Resource Files/Curriculum/advcluster/3158.txt (accessed 2011).

CNN. "World Population Projected to Reach 7 billion in 2011." August 12, 2009. articles.cnn.com/2009-08-12/tech/world.population_1_fertility-

rates-world-population-data-sheet-population-reference-bureau?_s=PM:
TECH (accessed 2011).

Diver, Steve. "Root Zone Heating for Greenhouse Crops." ATTRA (Appropriate
Technology Transfer for Rural Areas). April 2002. attra.ncat.org/attra-pub/
PDF/ghrootzone.pdf (accessed January 2011).

The End of the Line (documentary). Directed by Rupert Murray. 2009. Based on
the book by Charles Clover.

Friedman, Thomas L. *Hot, Flat and Crowded 2.0: Why We Need a Green
Revolution.* New York: Picador, 2009.

Hamre, Melvin L. "Farm Flock Poultry." University of Minnesota Extension.
2008. www.extension.umn.edu/distribution/livestocksystems/di3605.html
(accessed 2011).

Hill, Dennis R. and Terry Webster. "High pH Inhibits Nitrifying Bacteria."
Journal American Water Works Association, 2008: 20–22.

Hoffmann, M.P. and Frodsham, A.C. *Natural Enemies of Vegetable Insect Pests.*
Ithaca, NY: Cornell University, 1993.

Hoskin, Rebecca. "Permaculture — Farms for the Future." April 2, 2009. www.
viddler.com/explore/PermaScience/videos/4/ (accessed 2011).

Hughey, Travis. "The Barrel-ponics® Manual." Faith and Sustainable Technologies.
January 2011. www.fastonline.org/content/view/15/29/ (accessed December
2010).

Industries, Fritz. "Nitrifying Bacteria Facts." www.bioconlabs.com/
nitribactfacts.html (accessed December 2010).

Jolly, David. "Experts Debate Limits of Fish Farming." *New York Times.* January
31, 2011. www.nytimes.com/2011/02/01/science/earth/01fish.html?_r=3&
partner=rss&emc=rss (accessed February 1, 2011).

King, Ian. "Do It Yourself IBC System." *Backyard Aquaponics,* Q1 2011:
20–21.

Losinger, Willard C. "Feed-Conversion Ratio of Finisher Pigs in the USA." US
National Library of Medicine, National Institute of Health. July 11, 1995.
www.ncbi.nlm.nih.gov/pubmed/7584818 (accessed 2011).

McGinty, Andrew S. and James Rakocy. "Cage Culture of Tilapia." California
Aquaculture, University of California, Davis. aqua.ucdavis.edu/Database
Root/pdf/281FS.PDF (accessed 2010).

McWilliams, James E. *Just Food: Where Locavores Get It Wrong.* New York: Little,
Brown, 2009.

Mississippi State University Extension Service. "Commercial Catfish Production" MSUCares.com. October 14, 2010. http://msucares.com/aquaculture/catfish/ biology.html

National Science Academy. "Joint Science Academies' Statement on Growth and Responsibility." 2007. www.nationalacademies.org/includes/G8Statement_ Energy_07_May.pdf (accessed February 2011).

Nelson, Rebecca. "Soy, Barley, and Beer Show Promise in Fish Feed." *Aquaponics Journal*, Q1 2010: 20–21.

Penn State University. "Insecticidal Soap: An Alternative to Chemicals." Ken's Garden. January 2003. www.kensgardens.com/documents/tips/ insecticidal_soap.pdf (accessed 2011).

Pollen, Michael. *The Omnivore's Dilemma*. New York: The Penguin Press, 2006.

Save the Rainforest. "Facts About the Rainforest." Save the Rainforest. 2005. www.savetherainforest.org/savetherainforest_007.htm (accessed 2011).

Savidov, Nick. Ph D. "Evaluation and Development of Aquaponics Production and Product Market Capabilities." Final report, Brooks, AB: New Initiatives Fund, 2005.

Schoen, John W. "Global Food Chain Stretched to the Limit: Soaring Prices Spark Fears of Social Unrest in Developing World." MSNBC. January 14, 2011. www.msnbc.msn.com/id/41062817ns/business-consumer_news/ (accessed January 15, 2011).

Shelton, Anthony. PhD, Professor of Entemology, Cornell University. "Biological Control: A Guide to Nature's Enemies in North America." www.biocontrol. entomology.cornell.edu/predatorsTOC.html (accessed 2011).

Smith, Gar. "A Harvest of Heat: Agribusiness and Climate Change." Agribusiness Action Initiatives. Spring 2010. www.agribusinessaction.org/clearinghouse/ documents/AgriBizClimate4-8short.pdf (accessed 2011).

Timmons, M.B., Ebeling, J.M., 2010. *Recirculating Aquaculture (2nd Edition)*. Cayuga Aqua Ventures, LLC, Ithaca, N.Y.

USDA. "Irrigation and Water Use." USDA Briefing Room. November 22, 2004. www.ers.usda.gov/Briefing/WaterUse/ (accessed 2010).

US Environmental Protection Agency. "Integrated Pest Management (IPM) Principles." www.epa.gov/pesticides/factsheets/ipm.htm (accessed January 2011).

US Environmental Protection Agency. "Protecting Water Quality from Agricultural Runoff." Environmental Protection Agency. March 2005. water. epa.gov/polwaste/nps/upload/2005_4_29_nps_Ag_Runoff_Fact_Sheet.pdf (accessed 2011).

UN News Center. "Majority of World Population Face Water Shortages Unless Action Taken, Warns Migiro." February 5, 2009. www.un.org/apps/news/ story.asp?NewsID=29796&Cr=water&Cr1=agriculture (accessed 2011).

University of Michigan. "Human Appropriation of the World's Fresh Water Supply." University of Michigan Global Change 2. January 4, 2006. www.globalchange.umich.edu/globalchange2/current/lectures/freshwater_ supply/freshwater.html (accessed 2011).

VanGorder, Steven D. *Small Scale Aquaculture*. Breinigsville, PA: The Alternative Aquaculture Association, 2000.

Wheaton, Fred. Professor and Chairman, Dept. of Biological Resources Engineering. *Recirculating Aquaculture Systems: An Overview of Waste Management*. Academic findings, College Park, MD: University of Maryland.

World Wildlife Fund. *Living Planet Report*. Gland, Switzerland: WWF — World Wide Fund For Nature, 2008.

Worm, Kally. "Groundwater Drawdown." University of Wisconsin, Water is Life. Spring 2004. academic.evergreen.edu/g/grossmaz/WORMKA/ (accessed 2011).

Index

About the Author

For the past eight years Sylvia Bernstein's personal and professional lives have been centered on hydroponic and, more recently, aquaponic gardening. She is currently the President and Founder of The Aquaponic Source, one of a handful of US-based businesses focused entirely on the home aquaponic gardener. She also runs AquaponicsCommunity.com, a large online community site dedicated to aquaponic gardening, and writes the Aquaponic Gardening Blog, which is widely considered the most influential aquaponics blog in the world today. She also writes about aquaponics for *Urban Garden* and *Growing Edge* online magazines.

Author in her greenhouse.

In her pre-aquaponics life, Sylvia was the VP of Marketing, Innovation and Product Development for Aero-Grow International, the makers of the AeroGarden (a compact hydroponic garden designed for kitchen countertops). She was one of the company's original founders and was instrumental in developing the plant growth technology. She has a degree in agricultural economics from the University of California, Davis, and an MBA from the University of Chicago.

She lives with her husband, teenage son and Luna, a Tibetan terrier, in Boulder, Colorado (her daughter is away at college). Her inspiration is a large, thriving aquaponic setup in her backyard greenhouse in Boulder powered by tilapia, goldfish, koi and other creatures-that-swim.

Contact Information

Websites	AquaponicGardening.com TheAquaponicSource.com
Email	Sylvia@TheAquaponicSource.com
Community	AquaponicsCommunity.com
Blog	AquaponicGardeningBlog.com
Facebook	facebook.com/AquaponicGardening
Twitter	@aquapon

If you have enjoyed *Aquaponic Gardening* you might also enjoy other

BOOKS TO BUILD A NEW SOCIETY

Our books provide positive solutions for people who want to make a difference. We specialize in:

Sustainable Living • Green Building • Peak Oil • Renewable Energy
Environment & Economy • Natural Building & Appropriate Technology
Progressive Leadership • Resistance and Community
Educational & Parenting Resources

For a full list of NSP's titles, please call 1-800-567-6772 *or check out our website* at:

www.newsociety.com

NEW SOCIETY PUBLISHERS
Deep Green for over 30 years